Simple Solutions.

Minutes a Day–Mastery for a Lifetime!

Intermediate B Mathematics

Nancy L. McGraw

Bright Ideas Press, LLC
Cleveland, OH

Simple Solutions Intermediate B
Mathematics

Printed in the United States of America

ISBN: 978-1-934210-10-9

United States coin images from the United States Mint.

Cover Design: Dan Mazzola
Editor: Randy Reetz

Welcome

Simple Solutions.

Minutes a Day–Mastery for a Lifetime!

Note to the Student:

This workbook will give you the opportunity to practice skills you have learned in previous grades. By practicing these skills each day, you will gain confidence in your math ability.

Using this workbook will help you understand math concepts more easily and for many of you, it will give you a more positive attitude toward math in general.

In order for this program to help you be successful, it is extremely important that you do every lesson every day. It is also important that you ask your teacher for help on the problems you don't understand or get wrong when checking your homework.

If you put forth the effort, we guarantee that Simple Solutions will help you excel in mathematics.

Please keep this book out of a landfill – recycle it.

Lesson #1

1. 45,263 + 29,577 = ?
2. Find the median of 64, 56, and 88.
3. Draw two congruent pentagons.
4. Find the average (mean) of 6, 10, 56, and 104.
5. It is 4:15 now. What time was it six hours ago?
6. 7,000 − 3,556 = ?
7. How many feet are in a mile?
8. Draw an obtuse angle.
9. 5,344 × 7 = ?
10. Which digit is in the thousands place in 386,419?
11. The answer to a subtraction problem is called the __________.
12. 3,656 ÷ 6 = ?
13. What is the name of this shape?

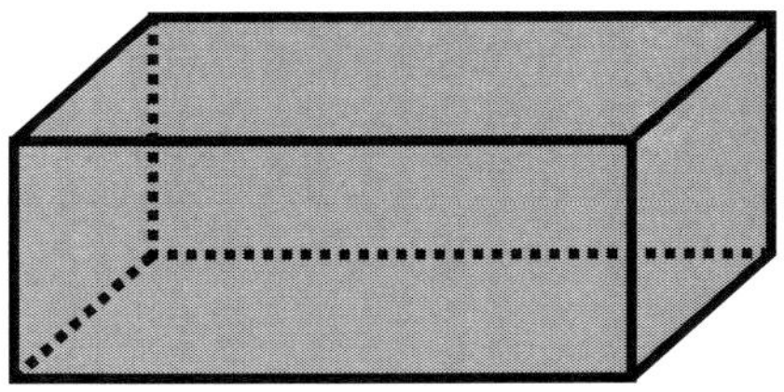

14. 45 × 24 = ?
15. How many dimes are in a dollar?
16. Round 56,213 to the nearest ten thousand.
17. Write the next number in the sequence. 46, 51, 56, …
18. How many cups are in a pint?
19. Janice, Helen, and Rita live in New York City, Chicago, or Detroit. Helen is the sister of the woman who lives in Chicago. Either Janice or Rita lives in Detroit. Janice lives alone. Which city does Janice live in? Is there enough information to solve this problem?
20. Write the definition for *median*.

1. 45,269 29,577 74,840	**2.** 64	**3.** 	**4.** 44 ×44 4 ⟌176 −16 016 −16 00
5. 10:15	**6.** 7,000 3,556 3,444	**7.** 5,280	**8.** 
9. 5344 × 7 37408	**10.** 6	**11.** difference	**12.** 608R8 6 ⟌3656 −36 0056 −48 08
13. rectangular Prism	**14.** 45 · 24 180 8100 8280	**15.** 10	**16.** 56,000
17. 61	**18.** 2	**19.** yes	**20.** the middle number

Lesson #2

1. How many quarts are in 5 gallons?
2. Draw perpendicular lines.
3. Jason's heart beats about 120 times per minute. How many times does Jason's heart beat in one day?
4. 263,455 + 198,663 = ?
5. Find the range of 86, 19, 25, 76, and 35.
6. Define *obtuse angle.*
7. 8,654 × 5 = ?
8. A six-sided polygon is called a(n) ________.
9. 306 ÷ 6 = ?
10. 703 − 159 = ?

11. The answer to a division problem is called the ____________.
12. Round 6,215,446 to the nearest thousand.
13. 3,762 ◯ 3,672
14. 66 × 51 = ?
15. 4.6 + 2.3 = ?
16. It is now 3:10. What time will it be 4 hours and 15 minutes from now?
17. The top number in a fraction is called the ___________.
18. $\frac{2}{5} + \frac{1}{5} = ?$
19. Put $\frac{4}{12}$ in simplest form.
20. Identify the type of angle.

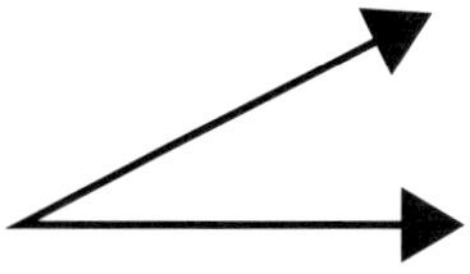

1.	2.	3.	4. 263455 + 198663 = 462118
5.	6. Large angle	7. 8654 × 5 = 43270	8. Pentagon
9. 306 ÷ 6 = 51	10. 703 − 159 = 544	11. quotient	12. 6,215,000
13. >	14. 66 × 51: 66 + 3300 = 3366	15. 4.6 + 2.3 = 6.9	16. 7:25
17. numerator	18. $\frac{3}{5}$	19. $\frac{2}{6}$	20. acute angle

Lesson #3

1. How many cups are in 5 pints?
2. $45 \times 23 = ?$

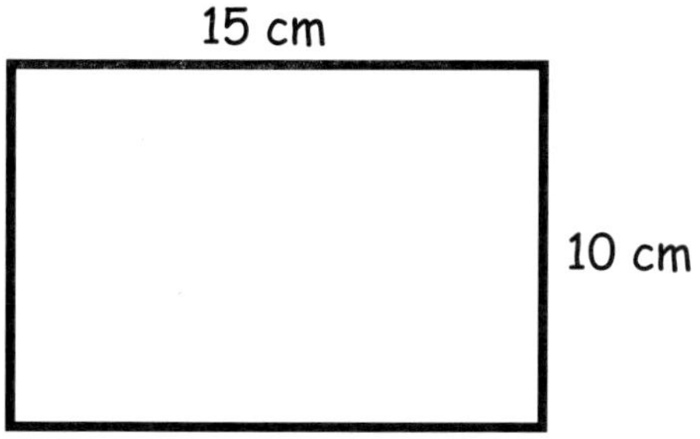

3. Draw intersecting lines.
4. $3{,}466 + 9{,}512 = ?$
5. Find the perimeter of this figure.
6. Put $\frac{5}{15}$ in simplest form.
7. Which digit is in the hundred thousands place in 365,217?
8. How many grams are in a kilogram?
9. How many degrees are in a right angle?

10. $60{,}000 - 25{,}415 = ?$
11. $465 \div 25 = ?$
12. $\frac{4}{5} = \frac{?}{15}$
13. $56 + 24 + 81 = ?$
14. Two decades before 1952 was what year?
15. If the diameter of a circle is 12 cm, what is its radius?
16. In one week a sapling's height grew from 45 inches to 63 inches. If the sapling continues to grow the same amount each week, how tall will it be three weeks after the first measurement?
17. Find the average (mean) of 54, 72, 90, and 12.
18. Draw a pentagon.
19. What is the bottom number in a fraction called?
20. $67.3 - 29.8 = ?$

1.	2.	3.	4.
5.	6.	7.	8.
9.	10.	11.	12.
13.	14.	15.	16.
17.	18.	19.	20.

Lesson #4

1. $5{,}490 \div 3 = ?$
2. How many ounces are in 3 pounds?
3. Round 2,416,304 to the nearest million.
4. $\frac{4}{9} - \frac{2}{9} = ?$
5. Find the median of 36, 14, 21, 56, and 10.
6. $957 \div 78 = ?$
7. Write 6.2 using words.
8. How many yards are in a mile?
9. A closed figure made up of line segments is called a(n) ____________.
10. $7{,}280 \times 9 = ?$
11. Find the perimeter of the triangle shown here.

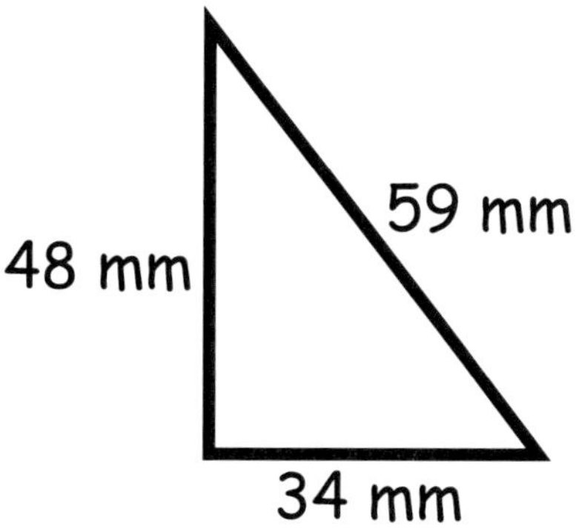

12. $4{,}216 - 1{,}775 = ?$
13. It is 2:00 now. What time was it 4 hours and 15 minutes ago?
14. Put $\frac{6}{9}$ in simplest form.
15. $4.7 - 2.9 = ?$
16. $\frac{4}{5} = \frac{8}{?}$
17. What fraction is shaded?

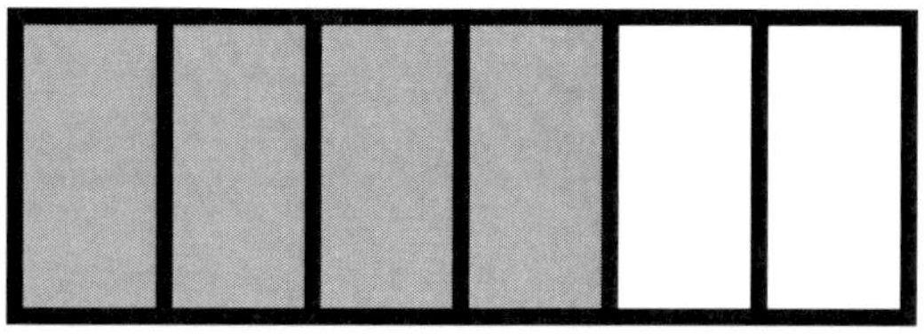

18. Draw an acute angle.
19. Jason and his family are spending 4 nights in Las Vegas. Their hotel costs $179.75 per night. How much should his family budget for hotel costs?
20. The answer in a multiplication problem is called the ____________.

1.	2.	3.	4.
5.	6.	7.	8.
9.	10.	11.	12.
13.	14.	15.	16.
17.	18.	19.	20.

Lesson #5

1. How many pints are in a quart?

2. $39 + 27 + 10 = ?$

3. $321 \times 25 = ?$

4. Draw a ray.

5. Write the next number in the sequence. 186, 179, 172, …

6. If the radius of a circle is 10 mm, what is the diameter?

7. Round 875,233 to the nearest ten thousand.

8. Find the mean of 160, 121, and 151.

9. $14{,}580 \div 9 = ?$

10. Draw a right angle.

11. $500 - 163 = ?$

12. Monica is putting candy into bags for the festival. She has 1,525 pieces of candy; there will be 25 pieces in each bag. How many bags of candy will Monica be able to make? Will she have any pieces of candy leftover?

13. A four-sided polygon is called a(n) _____________.

14. Define *line of symmetry.*

15. Find the area of the square to the right.

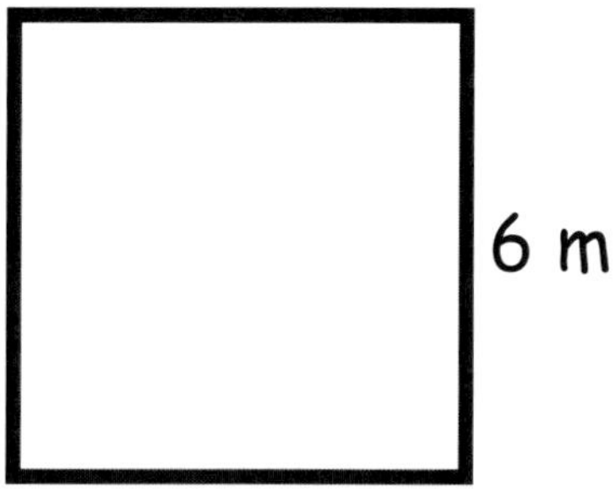

16. Put $\frac{12}{15}$ in simplest form.

17. Ninety years are _________ decades.

18. $86.25 - 17.6 = ?$

19. $\frac{4}{7} + \frac{3}{7} = ?$

20. Write *nine and twenty-three hundredths* as a decimal number.

1.	2.	3.	4.
5.	6.	7.	8.
9.	10.	11.	12.
13.	14.	15.	16.
17.	18.	19.	20.

Lesson #6

1. $\frac{5}{7} = \frac{?}{35}$
2. The sum is the answer to a(n) ____________ problem.
3. The top number in a fraction is called the ____________.
4. Draw a line segment.
5. Find the range of 98, 13, 41, 62, and 8.
6. Figures with the same size and shape are ____________.
7. Put $\frac{8}{12}$ in simplest form.
8. Write the name of the shape.

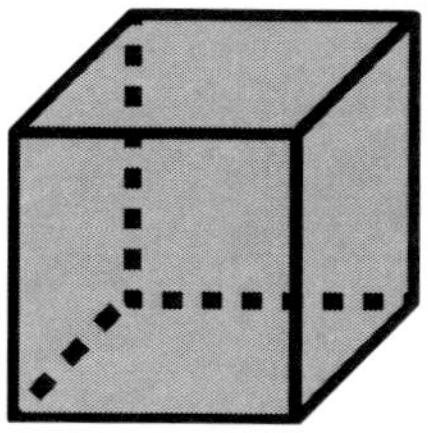

9. $14{,}237 \times 3 = ?$
10. Write 341,257 in expanded form.
11. Would a dime weigh about a gram or a kilogram?
12. $42{,}089 \div 8 = ?$
13. Six hundred years are how many centuries?
14. $900 - 328 = ?$
15. What is the probability of getting heads on one flip of a coin?
16. How many teaspoons are in 4 tablespoons?
17. $3.74 + 6.9 = ?$
18. $\frac{2}{9} + \frac{5}{9} = ?$
19. $374{,}219 + 455{,}667 = ?$
20. How many inches of snow fell in 1967? How much more snow fell in 1967 than in 1972?

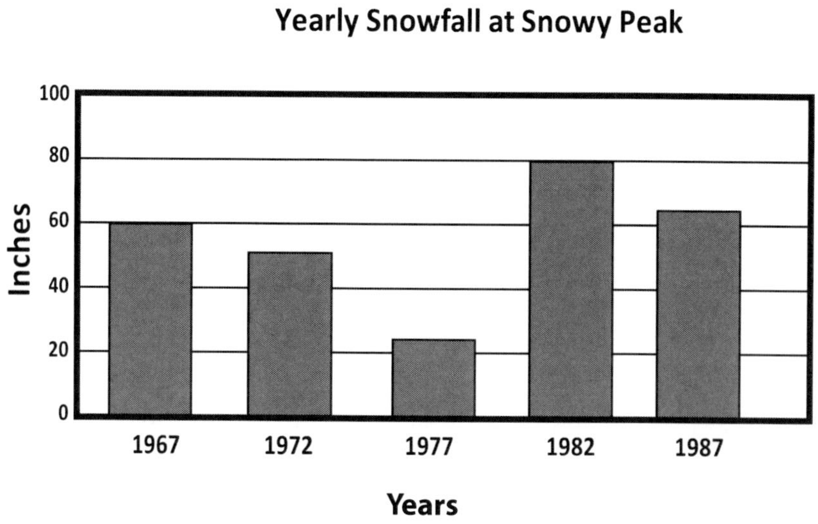

1.	2.	3.	4.
5.	6.	7.	8.
9.	10.	11.	12.
13.	14.	15.	16.
17.	18.	19.	20.

Lesson #7

1. Draw a horizontal line.
2. $396 \div 82 = ?$
3. What fraction is shaded?

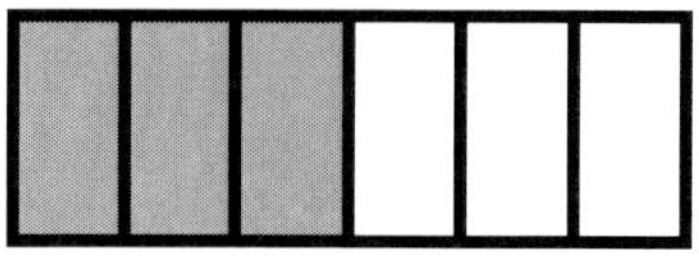

4. $516{,}677 + 78{,}918 = ?$
5. $12 \times 12 = ?$
6. Find the area.

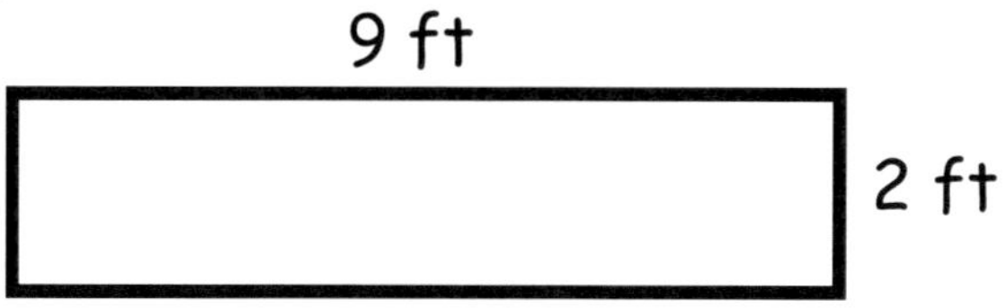

7. Write *six and four tenths* as a decimal number.
8. Put $\frac{5}{10}$ in simplest form.
9. Write $\frac{8}{6}$ as a mixed number.
10. $8{,}000 - 2{,}444 = ?$
11. Find the median of 16, 94, 66, 29, and 40.
12. If it is 12:40 now, what time will it be in 5 hours and 10 minutes?
13. How many yards are in a mile?
14. Draw two similar squares.

15. $\frac{8}{9} - \frac{3}{9} = ?$
16. $57 \div 7 = ?$
17. At the book fair, a small hardcover book costs \$12. A large hardcover book costs \$16, and a paperback costs \$7. If Terra began with \$30, and she has \$2 left, which two books did she buy?
18. Find the mean of 320, 294, 265, and 301.
19. $36 + \_\_\_ = 82$
20. Write the formula for finding the volume of a rectangular prism.

1.	2.	3.	4.
5.	6.	7.	8.
9.	10.	11.	12.
13.	14.	15.	16.
17.	18.	19.	20.

Lesson #8

1. How many inches are in a yard?
2. Write 7.23 using words.
3. How many minutes are in 4 hours?
4. The distance across a circle, through the center, is the ___________.
5. 3,127 − 1,769 = ?
6. Which digit is in the ten thousands place in 567,812?
7. A pentagon has ________ sides.
8. Draw perpendicular lines.
9. **There are a 1,000 mm in 1 meter.**
   Write *1,000 mm* = *1 m* three times.

10. 9.2 + 6.9 = ?
11. Put $\frac{4}{8}$ in simplest form.
12. 325 × 213 = ?
13. $\frac{8}{9} = \frac{?}{27}$
14. 3,467,893 + 5,784,229 = ?
15. Round 86,215 to the nearest hundred.
16. $\frac{7}{10} - \frac{4}{10} = ?$
17. What is the name of the shape to the right?

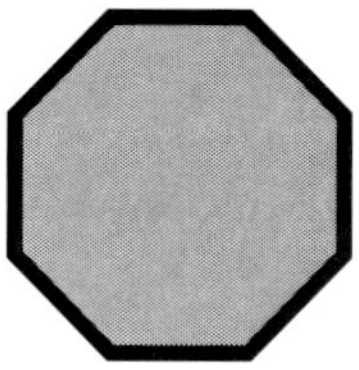

18. An angle less than 90° is called a(n) ________ angle.
19. There are 32 rows of cars on the lower level of a parking deck.
    If 28 cars are parked in each row, how many cars are on the lower level?
20. If a truck weighs 4 tons, how many pounds does it weigh?

1.	2.	3.	4.
5.	6.	7.	8.
9.	10.	11.	12.
13.	14.	15.	16.
17.	18.	19.	20.

Lesson #9

1. Find the area of the figure.

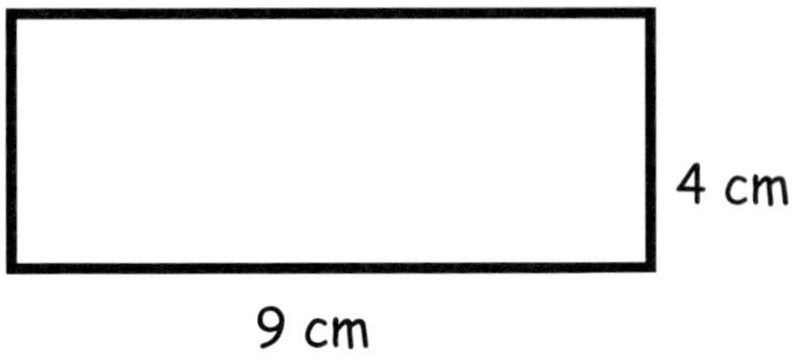

2. How much money is 9 quarters, 3 dimes, and 5 nickels?

3. 80,000 − 46,556 = ?

4. Is this figure a polygon?

5. Two figures having the same shape but with different sizes are ______.

6. 396 ÷ 82 = ?

7. Write the decimal number *seventeen and seven hundredths.*

8. 47,598 + 61,323 = ?

9. How many feet are in 4 yards?

10. 325 × 132 = ?

11. How many degrees are in a straight angle?

12. 75 − ___ = 36

13. What will be the time 8 minutes before 2:00?

14. At the start of week one, the odometer in Jan's van read 48,345 miles. By the end of week two, it read 50,525 miles. If she drove the same distance both weeks, how far did she drive each week?

15. Which digit is in the hundred thousands place in 2,671,835?

16. $\frac{7}{12} - \frac{5}{12} = ?$

17. How many millimeters are in a meter?

18. 18.23 + 7.6 = ?

19. Put $\frac{4}{16}$ in simplest form.

20. $\frac{4}{6} = \frac{?}{24}$

1.	2.	3.	4.
5.	6.	7.	8.
9.	10.	11.	12.
13.	14.	15.	16.
17.	18.	19.	20.

Lesson #10

1. 1,007 − 698 = ?

2. 1,407 ÷ 7 = ?

3. Classify the angle.

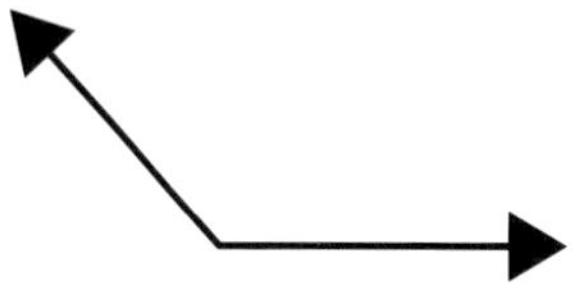

4. Find the mean of these numbers. 1,121 1,087 1,095

5. There are 228 first graders and 12 teachers. If each teacher has the same number of students, how many first graders are in each class?

6. Write 10.9 using words.

7. If the radius of a circle is 8 cm, what is its diameter?

8. 60 × 40 = ?

9. $\frac{2}{10} + \frac{3}{10} = ?$

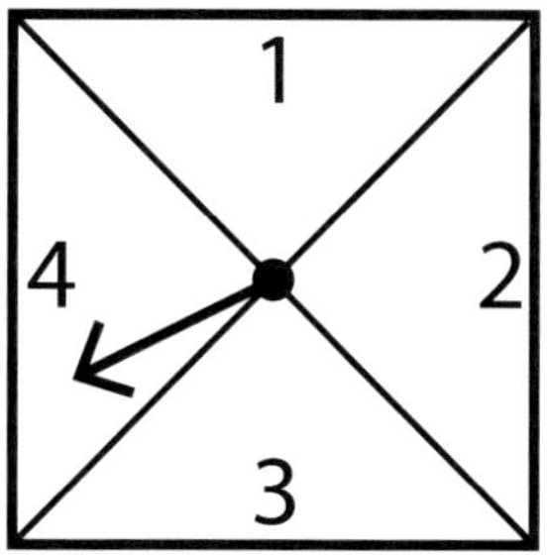

10. What is the probability of the spinner landing on a 3?

11. Nancy bought 8 golf balls for $24. At that price, what would she have to pay for 12 golf balls?

12. 32,804 × 3 = ?

13. Write $\frac{9}{7}$ as a mixed number.

14. Round 896,275 to the nearest hundred thousand.

15. How many months are in 3 years?

16. Give the name of this shape.

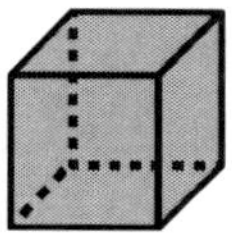

17. Draw intersecting lines.

18. Which is longer, 3 feet or 3 yards?

19. Put $\frac{10}{15}$ in simplest form.

20. Find the area of the square.

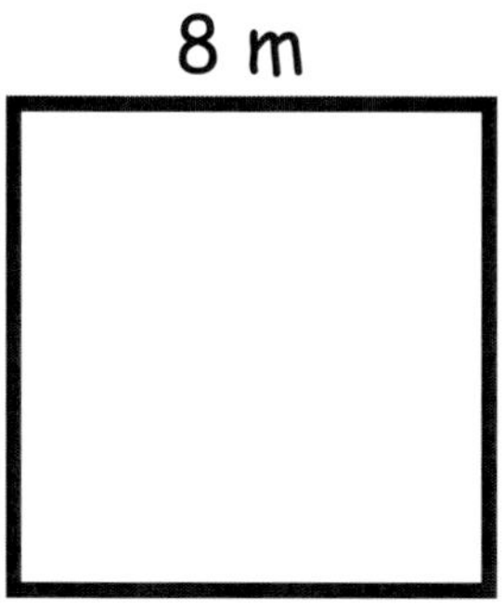

1.	2.	3.	4.
5.	6.	7.	8.
9.	10.	11.	12.
13.	14.	15.	16.
17.	18.	19.	20.

Lesson #11

1. The movie began at 7:30 and lasted 90 minutes. At what time did the movie end?

2. Write $6\frac{2}{5}$ as an improper fraction.

3. $60{,}275 + 24{,}845 = ?$

4. Draw a line segment.

5. $4{,}823 \div 23 = ?$

6. How many inches are in a yard?

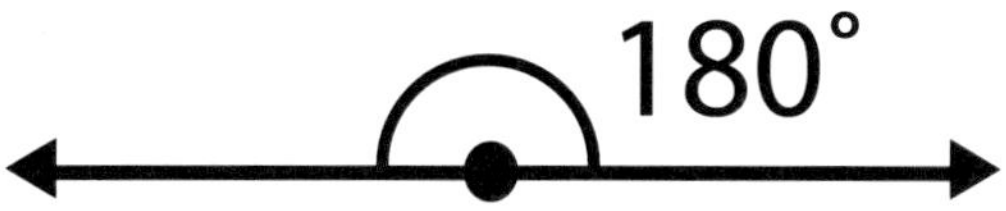

7. **A straight angle is a line. Its measure is 180°.** Draw a straight angle and label it 180°.

8. $125 \times 214 = ?$

9. Write $\frac{9}{5}$ as a mixed number.

10. $\frac{5}{8} = \frac{?}{16}$

11. $6.25 + 7.36 = ?$

12. Round 9,816,075 to the nearest million.

13. $\frac{5}{7} + \frac{2}{7} = ?$

14. $34{,}214 \times 7 = ?$

15. Draw a rectangle. Draw a line of symmetry through the rectangle.

16. $20 \times 3 - 15 = ?$

17. How many grams are in a kilogram?

18. $641 - 259 = ?$

19. Find the range of 42, 38, 14, and 66.

20. What do you call the number that occurs most often in a set of numbers?

1.	2.	3.	4.
5.	6.	7.	8.
9.	10.	11.	12.
13.	14.	15.	16.
17.	18.	19.	20.

Lesson #12

1. $\frac{1}{4} + \frac{1}{4} = ?$

2. Which is heavier, 6 grams or 6 kilograms?
3. Draw an obtuse angle.
4. 425 + 663 = ?
5. 2,884 ÷ 14 = ?
6. 96 × 36 = ?
7. Jake earns $8 per hour for cutting grass. If he works a 35-hour week, how much money will he earn?
8. Round 1,365,212 to the nearest ten thousand.
9. Any four-sided polygon is called a(n) ________.
10. What is $\frac{1}{2}$ of 60?
11. Put $\frac{3}{9}$ in simplest form.
12. What was the year two centuries before 1986?
13. Find the mean of 135, 375, and 450.
14. Write *three and twenty-seven hundredths* as a decimal.
15. $\frac{3}{4} = \frac{9}{?}$
16. Find the perimeter of this polygon.

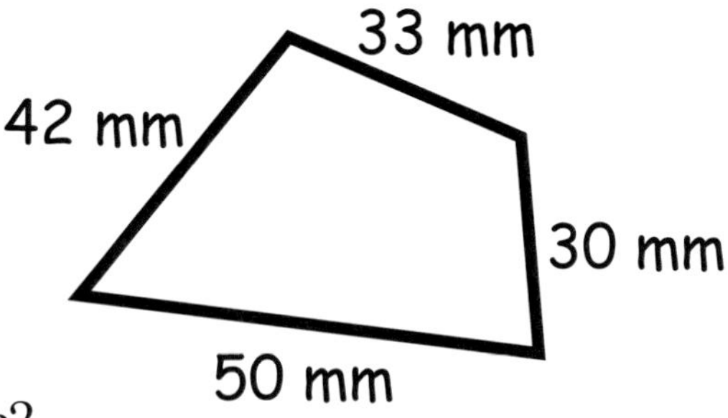

17. 5,050 − 1,385 = ?
18. How many degrees are in a straight angle?
19. There are ______ millimeters in a meter.
20. **On the Fahrenheit temperature scale, water freezes at 32°F, and it boils at 212°F.** Write *Fahrenheit freezing = 32°F* and *Fahrenheit boiling = 212°F.*

1.	2.	3.	4.
5.	6.	7.	8.
9.	10.	11.	12.
13.	14.	15.	16.
17.	18.	19.	20.

Lesson #13

1. Write $\frac{12}{5}$ as a mixed number.

2. $\frac{5}{8}-\frac{1}{8}=?$

3. Find the area.

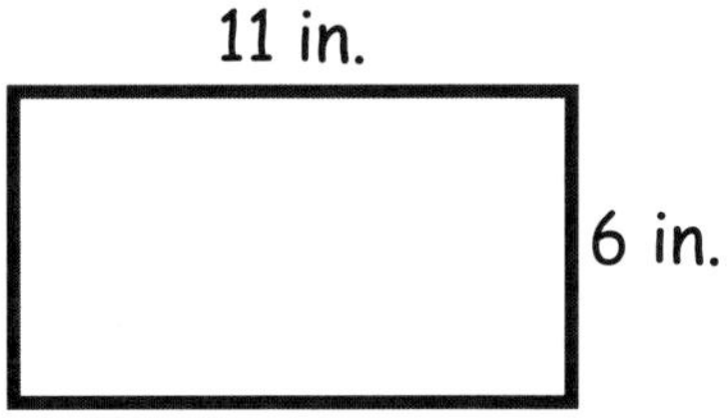

4. What is the probability of getting a 2 on one roll of a die?

5. On the Fahrenheit scale, water's freezing point is _______.

6. How many teaspoons are in 5 tablespoons?

7. Write *twelve and two hundredths* as a decimal.

8. 841,263 + 977,455 = ?

9. Round 8,314,205 to the nearest million.

10. Write $4\frac{1}{4}$ as an improper fraction.

11. $324 \times 213 = ?$

12. Are these shapes congruent?

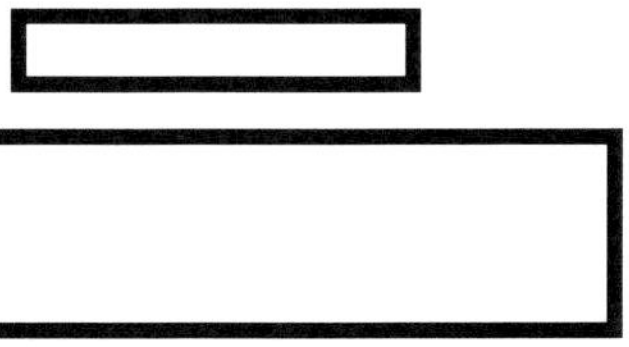

13. Ballet classes begin at 9:00 a.m. and repeat every 60 minutes. Jazz classes also begin at 9:00 a.m. and repeat every 45 minutes. What is the next time that both classes will begin at the same time?

14. 5,121 − 3,677 = ?

15. List the factors of 12.

16. How many decades are 40 years?

17. $8{,}452 \div 4 = ?$

18. Is 13 a prime or a composite number?

19. Which digit is in the ten millions place in 67,342,815?

20. Find $\frac{1}{5}$ of 25.

1.	2.	3.	4.
5.	6.	7.	8.
9.	10.	11.	12.
13.	14.	15.	16.
17.	18.	19.	20.

Lesson #14

1. Put $\frac{4}{6}$ in simplest form.

2. Draw a line of symmetry through this oval.

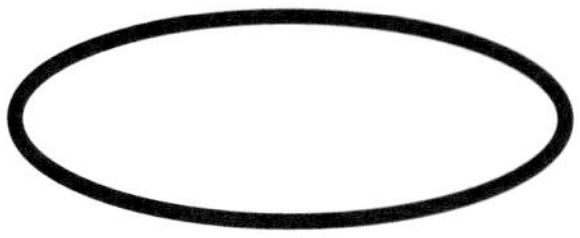

3. What will be the time 25 minutes after 6:00?

4. 43,204 − 19,561 = ?

5. Melanie made 25 scarves. She usually sells them for $18 each. If she sells 15 scarves at regular price and sells the remaining 10 scarves at half price, how much money will Melanie make?

6. 13.7 − 10.5 = ?

7. How many centimeters are in 3 meters?

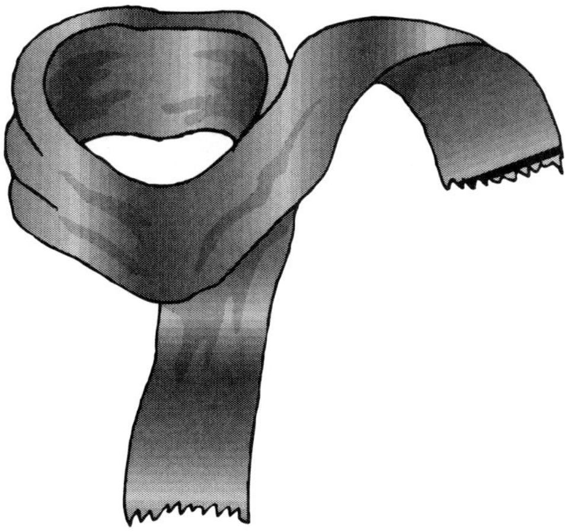

8. Is 24 a prime or a composite number?

9. 55 + 13 + 8 = ?

10. $\frac{9}{10} - \frac{2}{10} = ?$

11. 865 ÷ 25 = ?

12. 86 × 52 = ?

13. On the Fahrenheit temperature scale, water freezes at ________.

14. How many centuries are 600 years?

15. Find the perimeter of this figure.

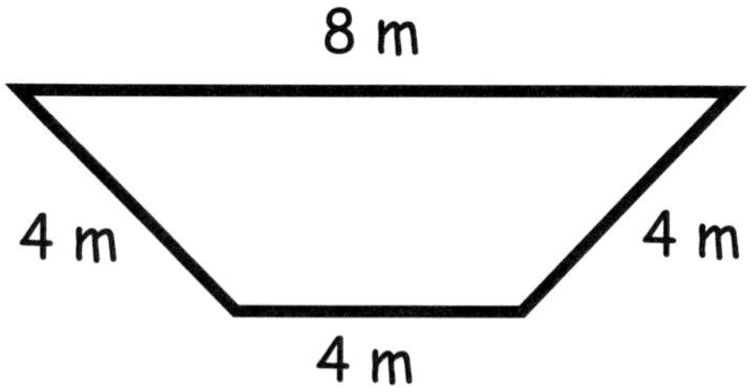

16. List the factors of 24.

17. $\frac{5}{9} = \frac{?}{36}$

18. Write the definition of *congruent.*

19. Find the mean of 75, 93, 88, and 60.

20. What kind of angle measures less than 90°?

1.	2.	3.	4.
5.	6.	7.	8.
9.	10.	11.	12.
13.	14.	15.	16.
17.	18.	19.	20.

Lesson #15

1. Write $\frac{14}{3}$ as a mixed number.
2. 633,215 + 45,866 = ?
3. How many degrees are in a straight angle?
4. Put $\frac{3}{12}$ in simplest form.
5. List the factors of 18.
6. 125 × 216 = ?
7. What is the probability of rolling a number greater than 4 on one roll of a die?
8. Find the perimeter of this figure.

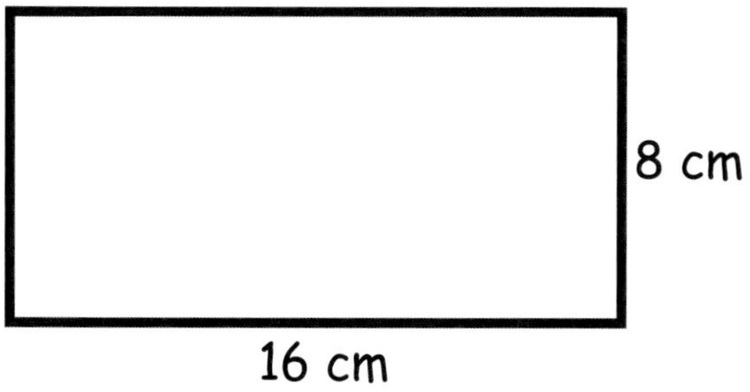

9. 27.9 − 13.2 = ?
10. Find the range of 37, 76, 53, and 25.
11. 720 ÷ 3 = ?
12. Round 6,414,862 to the nearest thousand.
13. Water boils at what temperature on the Fahrenheit scale?
14. Is the figure to the right a polygon?

15. Write 56,214 in expanded form.
16. 60,000 − 15,833 = ?
17. How many quarters are in $3.25?
18. If the radius of a circle is 12 meters, what is its diameter?
19. Draw a right angle. How many degrees are in a right angle?
20. Kim is traveling to China. She plans to take at least 400 pictures while she's there. If each roll of film contains 25 pictures, will 15 rolls of film be enough? Explain your answer.

1.	2.	3.	4.
5.	6.	7.	8.
9.	10.	11.	12.
13.	14.	15.	16.
17.	18.	19.	20.

Lesson #16

1. $8{,}036 \div 4 = ?$

2. What are the first four prime numbers?

3. Find the median of 37, 18, 56, 49, and 20.

4. $70 \times 30 = ?$

5. How many millimeters are in a meter?

6. List the factors of 14.

7. $302 - 179 = ?$

8. The answer to a subtraction problem is called the _________.

9. Write $\frac{11}{5}$ as a mixed number.

10. How many feet are in 2 miles?

11. $\frac{2}{6} + \frac{3}{6} = ?$

12. Define *composite numbers.*

13. $88{,}016 + 38{,}937 = ?$

14. Find the mean of 40, 94, and 106.

15. $\frac{5}{9} \bigcirc \frac{7}{8}$

16. Write $4\frac{2}{5}$ as an improper fraction.

17. If it is 9:30 now, what time will it be in 10 hours?

18. Put $\frac{3}{6}$ in simplest form.

19. How many DVDs were sold in 2004?

20. How many more DVDs were sold in 2006 than in 2005?

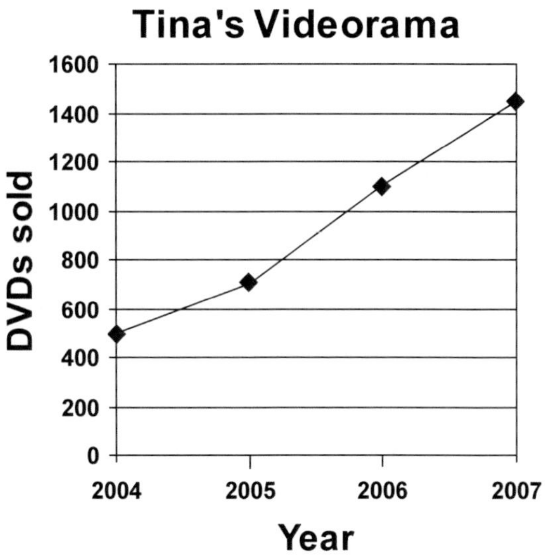

1.	2.	3.	4.
5.	6.	7.	8.
9.	10.	11.	12.
13.	14.	15.	16.
17.	18.	19.	20.

Lesson #17

1. $8.2 + 9.6 = ?$
2. Put $\frac{8}{16}$ in simplest form.
3. $36{,}842 - 19{,}689 = ?$
4. How many yards are in a mile?
5. Draw perpendicular lines.
6. Find the mean of 1,124, 723, and 859.
7. $295 \div 14 = ?$
8. Is 19 a prime number or a composite number? How do you know?
9. Find the perimeter of a pentagon if each side measures 6 inches.
10. $\frac{9}{10} \bigcirc \frac{3}{4}$
11. Two figures with the same shape but different sizes are _________.
12. List the factors of 15.
13. Write $\frac{6}{4}$ as a mixed number.
14. On the Fahrenheit scale, water freezes at ___________.
15. Nina and Patty made waffle batter for a school breakfast. They mixed 18 batches; each batch makes 12 waffles. How many waffles can they make in all?
16. Show a line of symmetry.

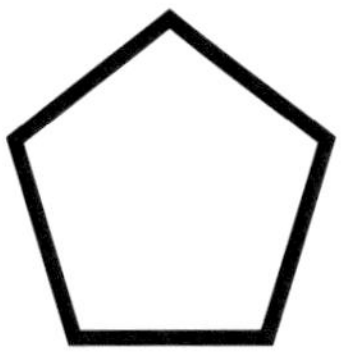

17. $\frac{3}{7} = \frac{?}{49}$
18. The quotient is the answer to a(n) _____________ problem.
19. Give the name of the shape.

20. $47 \times 93 = ?$

1.	2.	3.	4.
5.	6.	7.	8.
9.	10.	11.	12.
13.	14.	15.	16.
17.	18.	19.	20.

Lesson #18

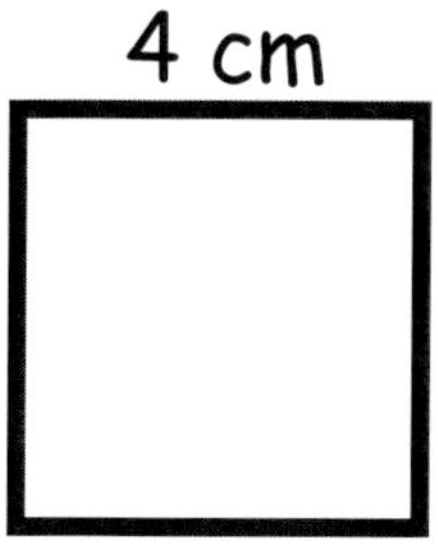

1. List the factors of 10.
2. What is the area of the square?
3. $\frac{3}{4} \bigcirc \frac{7}{8}$
4. Find the greatest common factor (GCF) of 10 and 15.
5. 97,316 + 29,407 = ?
6. Find the range of 81, 16, 43, 98, and 11.
7. Draw a ray.
8. How many millimeters are in 4 meters?
9. Is 25 a prime or a composite number?
10. 90,000 − 36,642 = ?
11. How many degrees are in a straight angle?
12. How many feet are in 4 yards?
13. 32 × 22 = ?
14. Define *radius*.
15. A hexagon has ________ sides.
16. Write the next number in the sequence. 234, 341, 448, …
17. Put $\frac{5}{10}$ in simplest form.
18. 4,535 ÷ 5 = ?
19. What was the total amount of money raised by both grades?
20. In which month did the eighth graders earn exactly twice as much as the seventh graders?

7th & 8th Grade Fundraisers

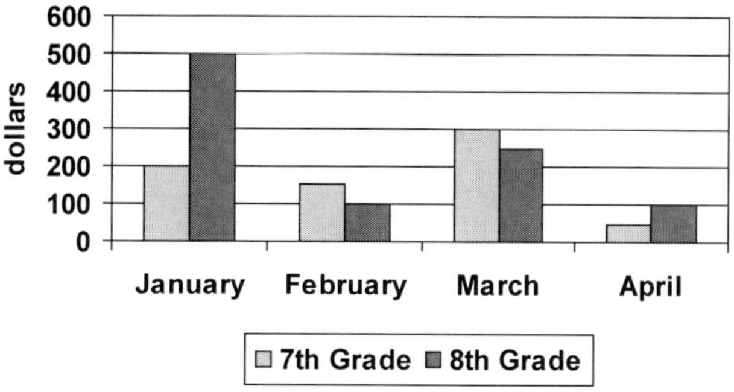

1.	2.	3.	4.
5.	6.	7.	8.
9.	10.	11.	12.
13.	14.	15.	16.
17.	18.	19.	20.

Lesson #19

1. $66 \times 23 = ?$

2. Define *prime number.*

3. Water boils at ______ degrees Fahrenheit.

4. $3{,}888 \div 36 = ?$

5. $1{,}006 - 599 = ?$

6. Write *sixteen and five tenths* as a decimal.

7. $34{,}175 \times 5 = ?$

8. Find the median of 65, 40, 25, 96, and 15.

9. $\frac{6}{7} - \frac{1}{7} = ?$

10. $55{,}819 + 7{,}349 = ?$

11. List the factors of 16.

12. Round 87,342,119 to the nearest million.

13. $\frac{3}{4} \bigcirc \frac{4}{5}$

14. How many cups are in 4 pints?

15. Find the greatest common factor (GCF) of 12 and 18.

16. Is 48 a prime number or a composite number?

17. How many ounces are in 6 pounds?

18. Put $\frac{12}{15}$ in simplest form.

19. Write $6\frac{1}{3}$ as an improper fraction.

20. A school bus carries a maximum weight of one ton. The average weight of each of the 18 students riding the bus is 95 pounds. The bus driver weighs 140 pounds. How many more students can ride the bus?

1.	2.	3.	4.
5.	6.	7.	8.
9.	10.	11.	12.
13.	14.	15.	16.
17.	18.	19.	20.

Lesson #20

1. The middle number in a set of numbers is the ____________.

2. 769,213 + 86,754 = ?

3. Give the probability of rolling a prime number on one roll of a die.

4. Find the area of the quadrilateral.

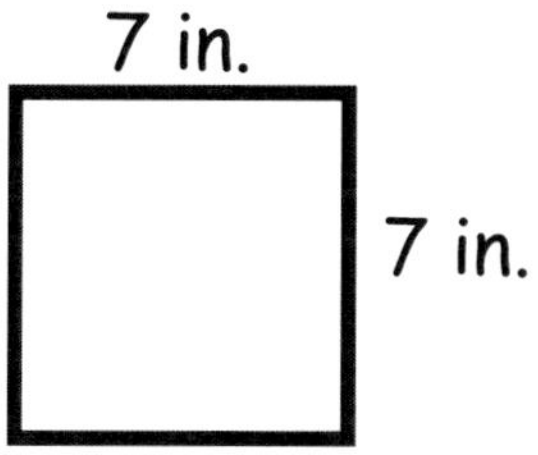

5. Draw perpendicular lines.

6. $331 \times 25 = ?$

7. 9,000 − 6,331 = ?

8. Find the least common multiple (LCM) of 12 and 16.

9. Which digit is in the hundred millions place in 815,367,402?

10. If it is 5:35 now, what time was it 4 hours and 15 minutes ago?

11. Find the mean of 320, 294, 265, and 301.

12. Put $\frac{5}{20}$ in simplest form.

13. An eight-sided polygon is called a(n) __________.

14. $5{,}760 \div 8 = ?$

15. How many feet are in 3 miles?

16. Are these figures similar?

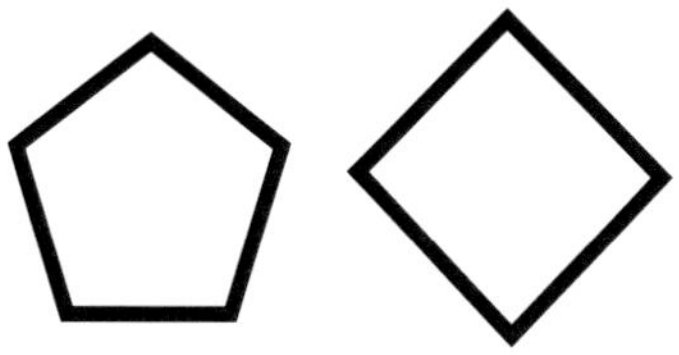

17. List the factors of 24.

18. How many weeks are 84 days?

19. $\frac{5}{8} = \frac{30}{?}$

20. Mitchell Elementary students are collecting soup labels to raise money for their school. In the first week, the students brought in 525 labels. They brought in 580 the second week and 635 labels in the third week. If the students collect soup labels at this rate, in which week will they bring in more than 1,000 labels?

1.	2.	3.	4.
5.	6.	7.	8.
9.	10.	11.	12.
13.	14.	15.	16.
17.	18.	19.	20.

Lesson #21

1. Find the GCF of 9 and 12.
2. 23,843 ÷ 4 = ?
3. List the factors of 14.
4. 46 + 24 + 13 = ?
5. A right angle measures _______ degrees.
6. Draw intersecting lines.
7. 85 × 53 = ?
8. How many years are 7 centuries?
9. What will be the time 10 minutes before 1:00?

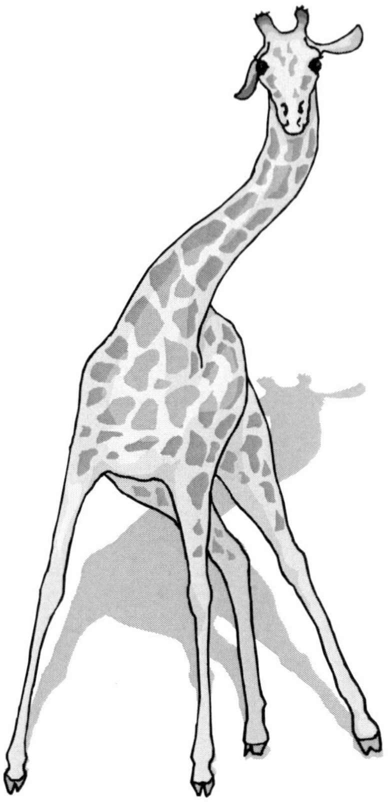

10. Find the perimeter.

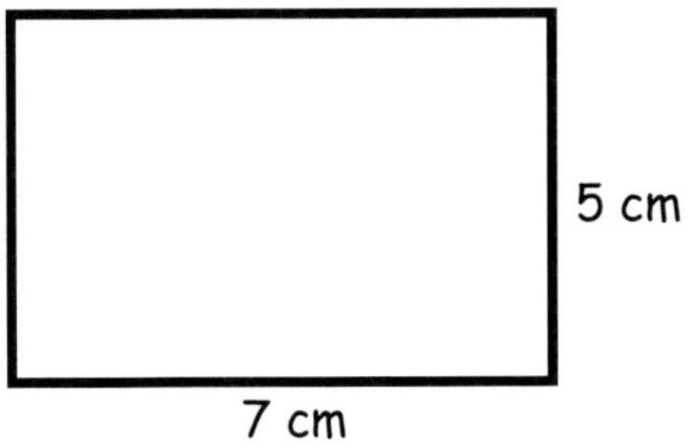

11. $\frac{5}{6} \bigcirc \frac{2}{3}$
12. Is 13 a prime or a composite number?
13. The answer to a multiplication problem is called the ___________.
14. Missy is using a 2 gallon punch bowl. She adds 2 pints of pineapple juice, 2 quarts of water, and 1 gallon of ginger ale. Is there enough room in the bowl for her to add 2 quarts of mango juice? Explain your answer.
15. How many dimes have the same value as $10?
16. How many pints are in a quart?
17. Round 7,887,530 to the nearest thousand.
18. Figures with the same size and shape are ___________.
19. Identify the type of angle.

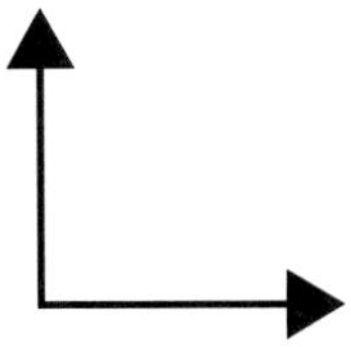

20. 4.6 − 1.7 = ?

1.	2.	3.	4.
5.	6.	7.	8.
9.	10.	11.	12.
13.	14.	15.	16.
17.	18.	19.	20.

Lesson #22

1. **On the Celsius scale, water freezes at 0°C and boils at 100°C.** Write *Celsius freezing = 0°C* and *Celsius boiling = 100°C.*

2. $\frac{9}{10} \bigcirc \frac{7}{8}$

3. $\frac{4}{5} + \frac{1}{5} = ?$

4. Draw a line segment.

5. Find the least common multiple (LCM) of 9 and 12.

6. How many millimeters are in 4 meters?

7. List the factors of 18.

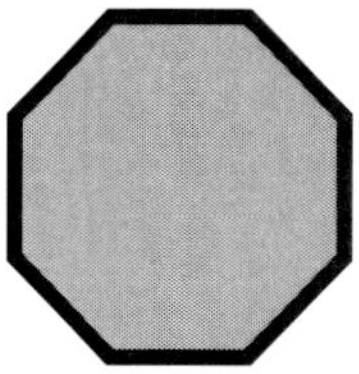

8. Identify this shape.

9. 48 + ___ = 83

10. Rick has 15 marbles. Lee has 3 times as many marbles as Rick. Jason has 8 more marbles than Lee. Hank has some marbles, too. If the four boys have a total of 167 marbles, how many marbles does Hank have?

11. Put $\frac{6}{18}$ in simplest form.

12. $12 \times 12 = ?$

13. If a baby anteater weighs 640 ounces, what is its weight in pounds?

14. If it is 2:55 now, what time was it 3 hours and 20 minutes ago?

15. Write 6.21 using words.

16. Half of a circle's diameter is called its _________.

17. $6{,}219 - 3{,}988 = ?$

18. How many feet are in 7 yards?

19. $973 \div 54 = ?$

20. Which is farther, 3 miles or 21,120 feet?

1.	2.	3.	4.
5.	6.	7.	8.
9.	10.	11.	12.
13.	14.	15.	16.
17.	18.	19.	20.

Lesson #23

1. 29,676 ÷ 4 = ?

2. Draw a ray.

3. A five-sided polygon is called a(n) ___________.

4. On the Celsius scale, water freezes at __________.

5. $\frac{5}{7} \bigcirc \frac{7}{8}$

6. Round 5,887,116 to the nearest hundred thousand.

7. 47 × 22 = ?

8. Find the median of 37, 19, and 50.

9. Which is longer, a meter or a kilometer?

10. 418,233 + 86,504 = ?

11. Are these shapes similar or congruent?

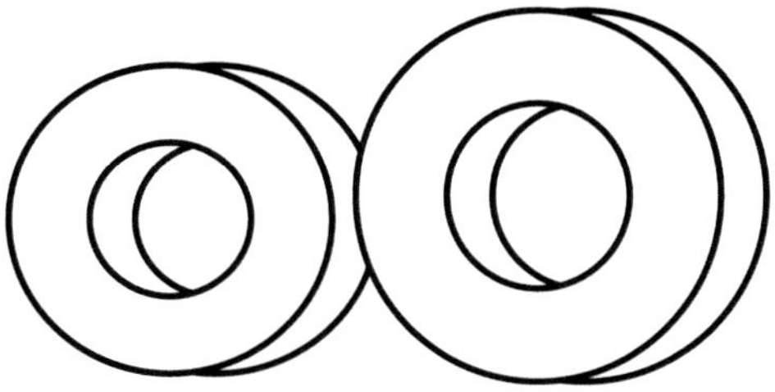

12. $\frac{5}{6} = \frac{?}{36}$

13. List the factors of 15.

14. Both Randy and Sue were at music lessons on Tuesday, May 3. Sue goes to music lessons every 4 days; Randy goes every 5 days. Lessons are Monday through Saturday. What is the next day of the week that they will both be at music lessons?

15. Which digit is in the millions place in 46,075,312?

16. Put $\frac{5}{25}$ in simplest form.

17. 60,000 − 28,199 = ?

18. Find the LCM of 8 and 15.

19. Write the first four prime numbers.

20. Find the perimeter of the square.

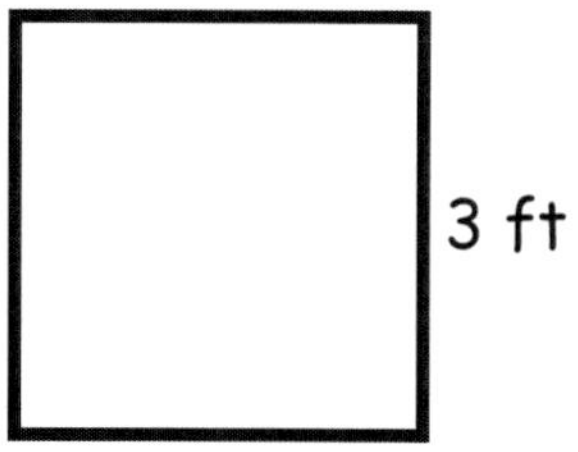

1.	2.	3.	4.
5.	6.	7.	8.
9.	10.	11.	12.
13.	14.	15.	16.
17.	18.	19.	20.

Lesson #24

1. Find the GCF of 15 and 25.

2. How many teaspoons are in 7 tablespoons?
3. Would a bucket of popcorn weigh about 900 grams or 900 kilograms?
4. $5 \times 5 \times 3 = ?$
5. Write 31,226 in expanded form.
6. $6.32 + 15.8 = ?$
7. Draw a line of symmetry.

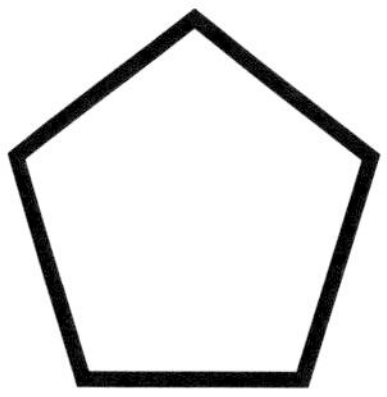

8. $\frac{3}{4} \bigcirc \frac{2}{3}$
9. On the Celsius scale, water boils at ______.
10. $455 \times 32 = ?$
11. How many yards are in 2 miles?
12. $4{,}102 - 1{,}667 = ?$
13. Sean is 4 feet, 7 inches tall. Paul is 59 inches tall. Who is taller?
14. List the factors of 20.
15. $72{,}095 \div 45 = ?$
16. Write $5\frac{2}{3}$ as an improper fraction.
17. What time will it be in 90 minutes, if it is 7:00 now?
18. Put $\frac{3}{15}$ in simplest form.
19. Draw parallel horizontal lines.
20. What is the probability that the spinner will land on a consonant? On the letter A?

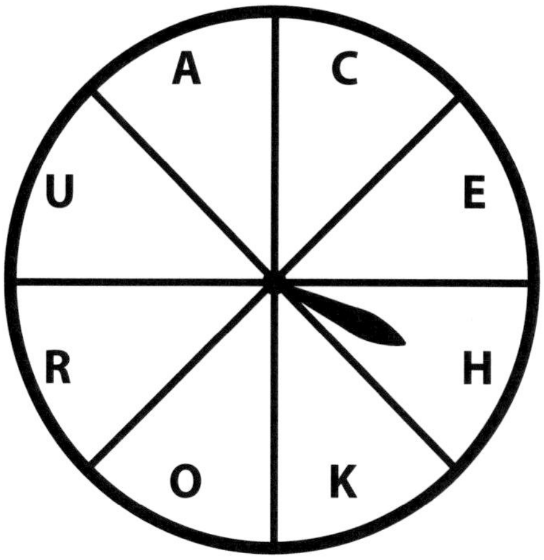

1.	2.	3.	4.
5.	6.	7.	8.
9.	10.	11.	12.
13.	14.	15.	16.
17.	18.	19.	20.

Lesson #25

1. Write $\frac{12}{7}$ as a mixed number.
2. Find the LCM of 16 and 18.
3. $503 - 175 = ?$
4. $\frac{3}{5} \bigcirc \frac{6}{7}$
5. Draw a hexagon.
6. $56{,}304 + 27{,}138 = ?$
7. Put $\frac{8}{12}$ in simplest form.

8. A school raised $950 for new gym shirts. If the school purchases 75 shirts, it will receive 5 shirts free. If each shirt costs $12, how many shirts will the school have, including any free ones?
9. List the factors of 25.
10. Find $\frac{1}{6}$ of 24.
11. Find the area of the rectangle.

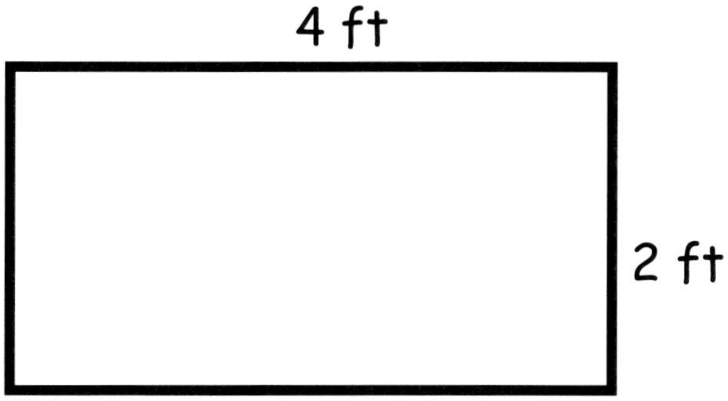

12. How many inches are 6 feet?
13. $15{,}345 \times 6 = ?$
14. How many grams are in 7 kilograms?
15. Closed figures made up of line segments are called ______.
16. $18{,}345 \div 3 = ?$
17. Give the Celsius and Fahrenheit freezing temperatures of water.
18. Round 374,816 to the nearest ten thousand.
19. Find the mode of 86, 39, 27, 39, and 80.
20. Make a factor tree for 24.

1.	2.	3.	4.
5.	6.	7.	8.
9.	10.	11.	12.
13.	14.	15.	16.
17.	18.	19.	20.

Lesson #26

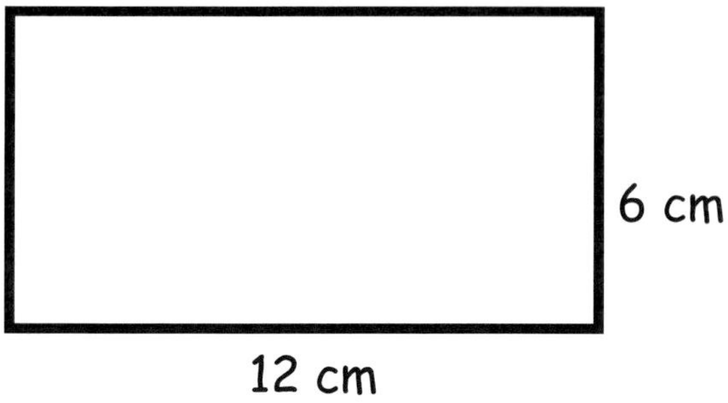

1. $20{,}000 - 9{,}335 = ?$
2. Find the area of the figure.
3. $\frac{8}{9} \bigcirc \frac{6}{7}$
4. Find $\frac{1}{7}$ of 49.
5. $6{,}429 + 8{,}552 = ?$
6. Draw an acute angle.
7. $15{,}345 \div 5 = ?$
8. Is 23 a prime or a composite number?
9. Which digit is in the thousands place in 7,815,042?
10. How many cups are in 3 pints?
11. $551 \times 32 = ?$
12. Write *twenty-five and three hundredths* as a decimal.
13. Find the LCM of 8 and 14.
14. $\frac{3}{4} = \frac{?}{20}$
15. The answer to an addition problem is called the _______.
16. Put $\frac{7}{35}$ in simplest form.
17. Marissa wants to buy a new cell phone that costs \$199. She works 13 hours a week after school. If Marissa is paid \$7 an hour, in how many weeks will she be able to afford to buy her cell phone?
18. List the factors of 20.
19. Are these lines parallel to each other?
20. Make a factor tree for 72.

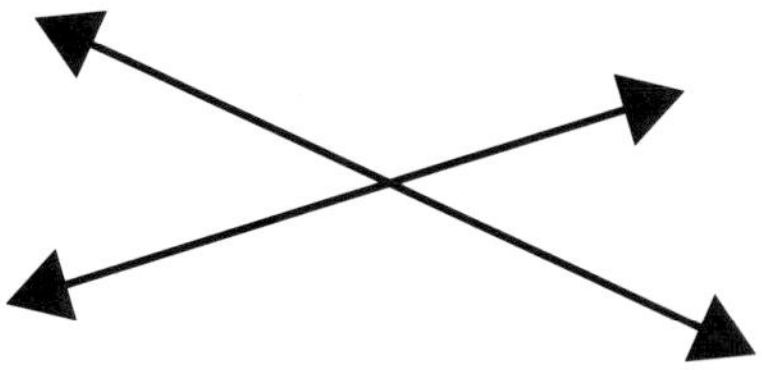

1.	2.	3.	4.
5.	6.	7.	8.
9.	10.	11.	12.
13.	14.	15.	16.
17.	18.	19.	20.

Lesson #27

1. Find the area.

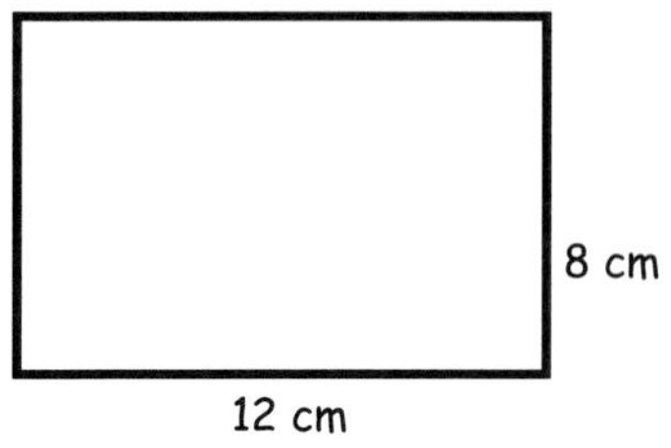

2. $\frac{8}{9} - \frac{3}{9} = ?$

3. Find the mode in 14, 25, 55, 14, and 62.

4. Find the median of the set of numbers in item 3.

5. Make a factor tree for 45.

6. 183,214 + 88,752 = ?

7. How many quarts are in 8 gallons?

8. $3{,}255 \times 4 = ?$

9. Draw a straight angle.

10. 20,000 − 6,431 = ?

11. Between 11:00 a.m. and 11:00 p.m. Hot Dog Heaven served 1,884 hotdogs. On average, how many hotdogs were served per hour?

12. Put $\frac{6}{24}$ in simplest form.

13. Write $\frac{13}{7}$ as a mixed number.

14. How many ounces are in 4 pounds?

15. At what temperature does water boil in degrees Fahrenheit?

16. Find the GCF of 14 and 21.

17. Write $7\frac{4}{5}$ as an improper fraction.

18. $650 \div 25 = ?$

19. 235.76 − 97.8 = ?

20. Are these shapes congruent or similar?

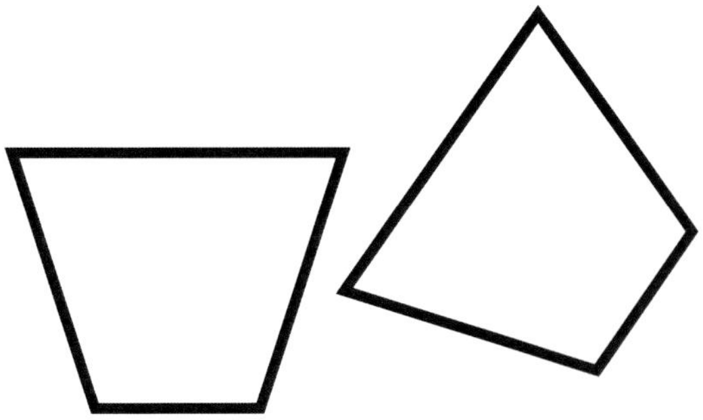

1.	2.	3.	4.
5.	6.	7.	8.
9.	10.	11.	12.
13.	14.	15.	16.
17.	18.	19.	20.

Lesson #28

1. The distance across the middle of a circle is called the _______.

2. $4 \times 28 = ?$

3. Draw two lines of symmetry.

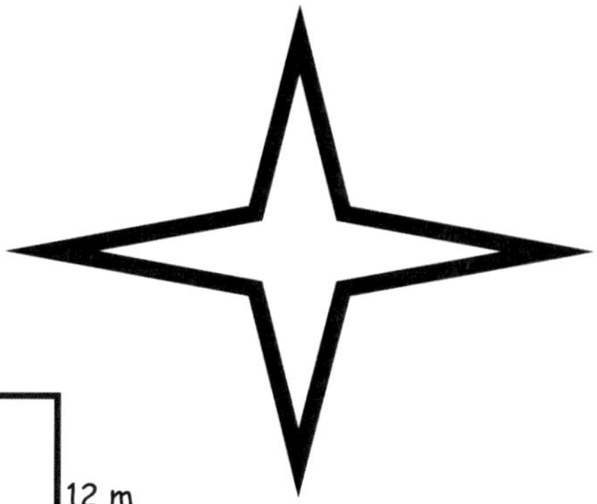

4. Find $\frac{1}{3}$ of 18.

5. Find the perimeter of the square.

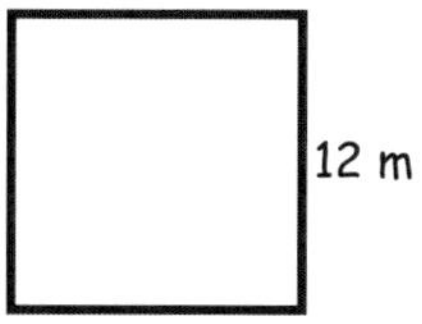

6. $23{,}513 - 18{,}658 = ?$

7. A shrub is 156 inches tall. What is the shrub's height in feet?

8. Round 32,415,266 to the nearest thousand.

9. Put $\frac{10}{15}$ in simplest form.

10. 28,604 ◯ 28,406

11. $281 + 48 = ?$

12. Make a factor tree for 12.

13. Find the mean of 51, 52, and 56.

14. $4{,}155 \div 5 = ?$

15. Which digit is in the ten thousands place in 25,740?

16. Write $5\frac{2}{5}$ as an improper fraction.

17. Identify the type of angle.

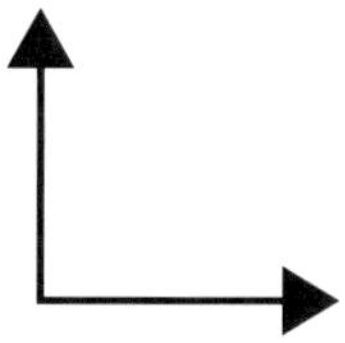

18. $\frac{2}{4}$ ◯ $\frac{4}{8}$

19. $9.8 + 6.3 = ?$

20. What is the probability of landing on an even number? On a number greater than 3?

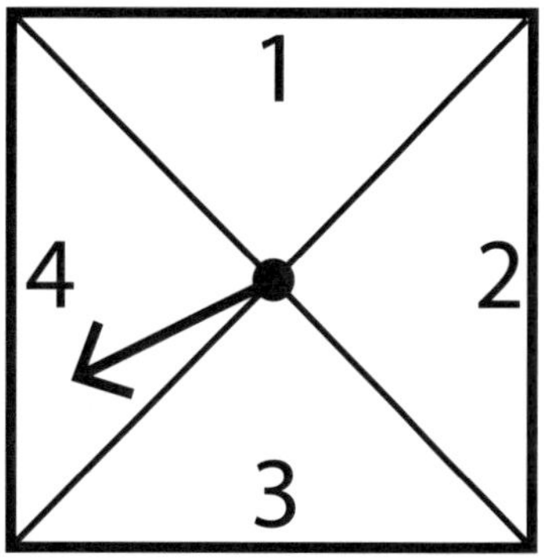

1.	2.	3.	4.
5.	6.	7.	8.
9.	10.	11.	12.
13.	14.	15.	16.
17.	18.	19.	20.

Lesson #29

1. Find the mean of 72, 53, 68, and 71.

2. Draw a ray.

3. $976 \div 6 = ?$

4. If the radius of a circle is 14 cm, what is its diameter?

5. $300 \times 600 = ?$

6. Put $\frac{25}{30}$ in simplest form.

7. List the factors of 21.

8. $4{,}000 - 885 = ?$

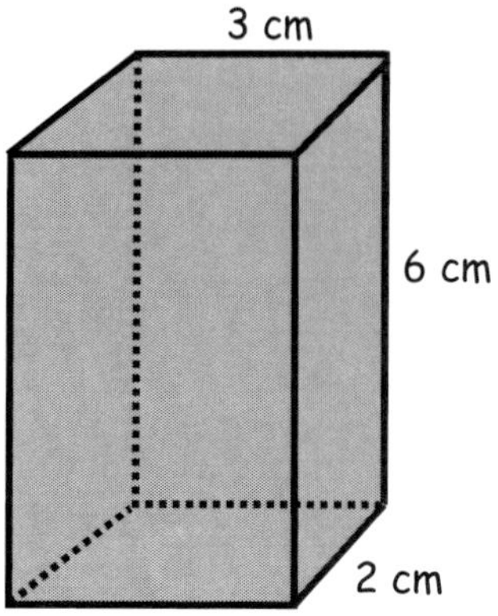

9. ***Volume* is a measure of space inside of a solid figure. The formula for finding the volume of a rectangular prism is length × width × height. ($V = l \times w \times h$)** Find the volume of the figure above.

10. Make a factor tree for 32.

11. What will be the time 80 minutes after 5:00?

12. $163{,}214 + 79{,}456 = ?$

13. $\frac{7}{10} \bigcirc \frac{8}{9}$

14. Find the GCF and the LCM of 12 and 16.

15. A five-sided polygon is called a(n) ________.

16. How many weeks are in 7 months?

17. Write the first 3 prime numbers.

18. $\frac{4}{7} = \frac{16}{?}$

19. Ken spent $25 on jeans and $32 on a sweater. Then his sister gave him $10. Now, Ken has $31. How much did he have before buying the jeans and sweater?

20. Write $\frac{12}{5}$ as a mixed number.

1.	2.	3.	4.
5.	6.	7.	8.
9.	10.	11.	12.
13.	14.	15.	16.
17.	18.	19.	20.

Lesson #30

1. Sean's bedroom is 10 feet long by 8 feet wide. He wants to cover the floor with tiles that are 1 foot long and 1 foot wide. How many tiles will he need?

2. List the Celsius and Fahrenheit boiling points of water.

3. Find the mode of 33, 59, 86, 24, and 59.

4. $83 \times 38 = ?$

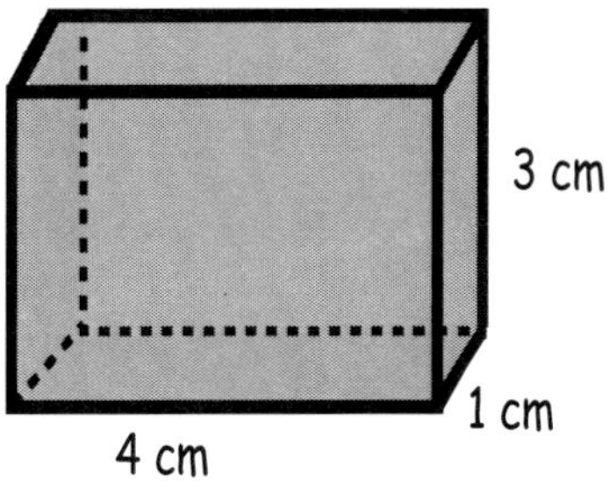

5. How many inches are in 3 feet?

6. $600{,}000 - 432{,}864 = ?$

7. Find the volume of this solid figure.

8. How many teaspoons are in 9 tablespoons?

9. Make a factor tree for 56.

10. $\frac{8}{9} + \frac{1}{9} = ?$

11. How many pints are in 8 quarts?

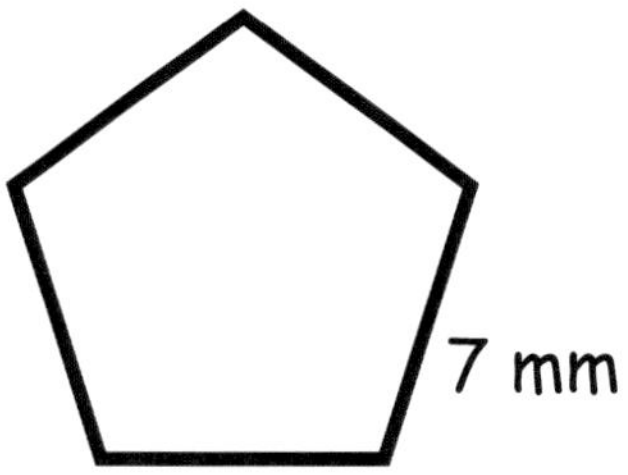

12. $864 + 399 = ?$

13. Find the perimeter of the pentagon.

14. It is 1:25 now. What time will it be 6 hours and 20 minutes from now?

15. An angle that measures more than 90° is called __________.

16. Put $\frac{4}{8}$ in simplest form.

17. $3{,}963 \div 36 = ?$

18. Which is longer, 8 yards or 8 feet?

19. Find the range of 100, 92, 87, and 51.

20. Write $7\frac{1}{3}$ as an improper fraction.

1.	2.	3.	4.
5.	6.	7.	8.
9.	10.	11.	12.
13.	14.	15.	16.
17.	18.	19.	20.

Lesson #31

1. Find $\frac{1}{3}$ of 27.

2. List the factors of 24.

3. Draw a line segment.

4. Find the volume of the cube.

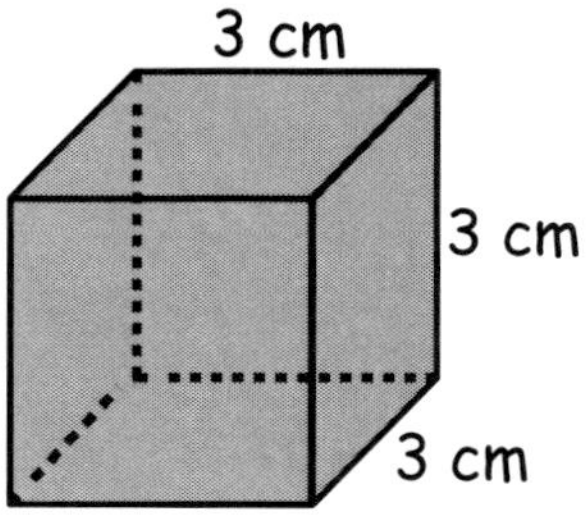

5. $40 \times 37 = ?$

6. Is the number 36 prime or composite?

7. $7.3 - 5.7 = ?$

8. $63 + \_\_\_ = 74$

9. Put $\frac{10}{25}$ in simplest form.

10. It is 3:05. What time was it 4 hours ago?

11. On the Celsius scale, water freezes at _________.

12. How many minutes are in 5 hours?

13. Kyle works at a barber shop on Thursday, Saturday, and Monday for a total of 16 hours. He works twice as many hours on Saturday as on Thursday. He works 4 hours on Monday. How many hours does Kyle work on Saturday?

14. $363{,}225 + 88{,}726 = ?$

15. $8{,}346 \div 6 = ?$

16. Draw perpendicular lines.

17. Make a factor tree for 40.

18. What is the GCF of 15 and 20?

19. Write $5\frac{2}{3}$ as an improper fraction.

20. What is the probability of getting heads on one flip of a coin?

1.	2.	3.	4.
5.	6.	7.	8.
9.	10.	11.	12.
13.	14.	15.	16.
17.	18.	19.	20.

Lesson #32

1. Any four-sided polygon is called a(n) _________.

2. Put $\frac{14}{16}$ in simplest form.

3. Round 16,244,336 to the nearest ten thousand.

4. $37 \times 81 = ?$

5. List the factors of 18.

6. $46.3 + 10.75 = ?$

7. Find the LCM of 15 and 18.

8. Fifty years are ______ decades.

9. $137{,}216 + 886{,}774 = ?$

10. Closed figures made up of line segments are called ________.

11. Find the mode in 86, 41, 32, 41, and 75.

12. Draw a straight angle. How many degrees are in a straight angle?

13. $40{,}000 - 18{,}215 = ?$

14. The answer to a division problem is the ___________.

15. How many inches are in 4 feet?

16. Find the volume of the prism.

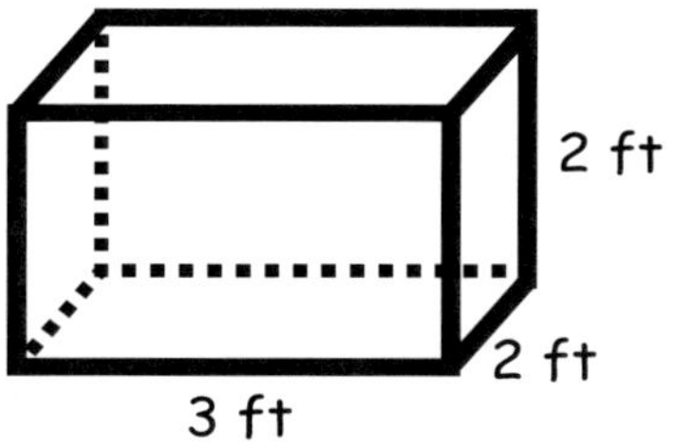

17. $\frac{7}{9} - \frac{2}{9} = ?$

18. Numbers with only two factors are _______ numbers.

19. $6{,}543 \div 3 = ?$

20. Ben, Elliot, Henry, and Alex are in line for the roller coaster. Ben is third in line. Henry is not first or second; Alex in not first. In what position is each person?

1.	2.	3.	4.
5.	6.	7.	8.
9.	10.	11.	12.
13.	14.	15.	16.
17.	18.	19.	20.

Lesson #33

1. Find the GCF of 12 and 18.
2. Write *seventeen and one hundredth* as a decimal.
3. Write the first five prime numbers.
4. $\frac{7}{15} - \frac{1}{15} = ?$
5. How many degrees are in a right angle?
6. Write $6\frac{2}{3}$ as an improper fraction.
7. 267,314 + 562,471 = ?
8. Find the mean and the mode of 271, 298, and 271.
9. 37,650 ÷ 5 = ?
10. If the radius of a circle is 8 inches, what is its diameter?
11. What is the name of this type of shape?

12. 348 × 24 = ?
13. How many weeks are 105 days?
14. Round 756,219,045 to the nearest million.
15. 302 − 89 = ?
16. An angle that measures less than 90° is called ___________.
17. Write $\frac{17}{5}$ as a mixed number.
18. Find the median of 18, 37, 96, 55, and 23.
19. Write the formula for finding the volume of a rectangular solid.
20. Columbus, Ohio is about 786 miles from Omaha, Nebraska. If a train makes 3 round trips from Columbus to Omaha, how far does it travel?

1.	2.	3.	4.
5.	6.	7.	8.
9.	10.	11.	12.
13.	14.	15.	16.
17.	18.	19.	20.

Lesson #34

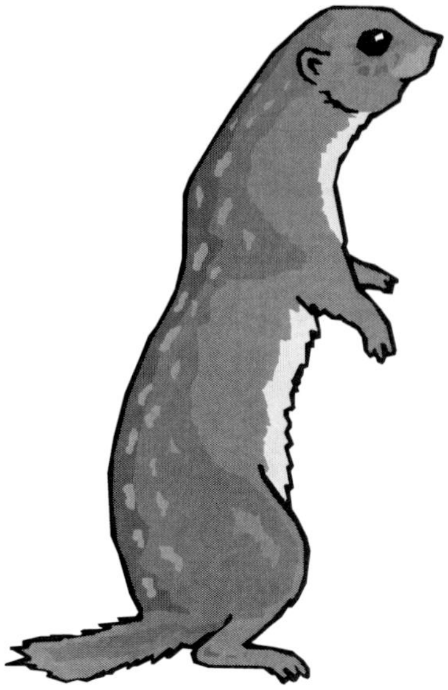

1. $\frac{9}{16} - \frac{5}{16} = ?$
2. Find the LCM of 6 and 15.
3. Which is greater, 300 feet or 80 yards?
4. Find $\frac{1}{7}$ of 56.
5. 75,000 − 26,312 = ?
6. Put $\frac{3}{18}$ in simplest form.
7. How many feet are in 4 miles?
8. Write $\frac{19}{7}$ as a mixed number.
9. A five-sided polygon is called a(n) _________.
10. Find the volume of the prism to the right.
11. List the factors of 14.

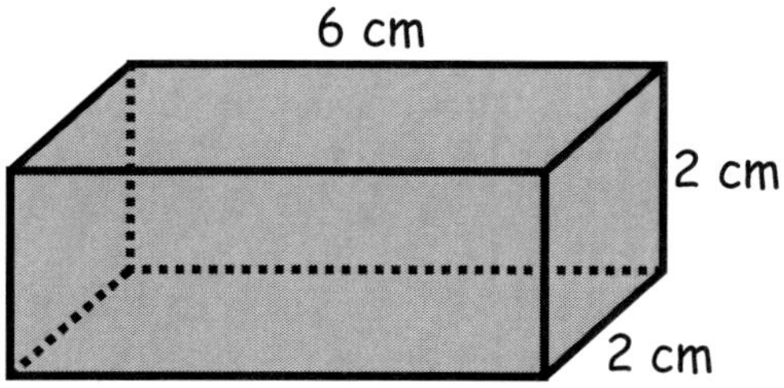

12. Marvin bikes an average of 13 kilometers per day. About how many kilometers does Marvin bike in 3 weeks?
13. Which digit is in the hundred thousands place in 46,780,213?
14. How many cups are in 5 pints?
15. The distance across a circle, through the center, is the _________.
16. 685 ÷ 35 = ?
17. 44 + 16 + 28 = ?
18. Find the range of 103, 119, 110, 152, and 174.
19. 25,614 × 4 = ?
20. Find the perimeter of a rectangle whose length is 7 cm and whose width is 4 cm.

1.	2.	3.	4.
5.	6.	7.	8.
9.	10.	11.	12.
13.	14.	15.	16.
17.	18.	19.	20.

Lesson #35

1. Define the term *similar.*
2. How many centimeters are in 6 meters?
3. Put $\frac{4}{24}$ in simplest form.
4. $93 \times 37 = ?$
5. If it is 4:45 now, what time was it 3 hours and 10 minutes ago?
6. The answer to a multiplication problem is the ____________.
7. Draw perpendicular lines.
8. Find the probability of rolling an odd number on one roll of a die.
9. $\frac{5}{7} = \frac{?}{21}$
10. How many quarters are in $6.00?
11. Which activity was the girls' favorite?
12. About how many more boys play video games compared to sending email to their friends?

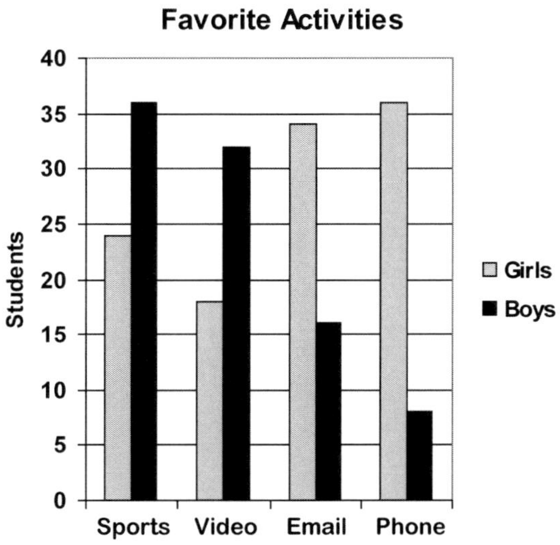

13. $3.6 + 4.9 = ?$
14. Write $3\frac{2}{3}$ as an improper fraction.
15. Would yogurt best be measured in milliliters or liters?
16. Find the GCF of 12 and 18.
17. How many feet are in 6 yards?
18. Write 15.9 using words.
19. $41{,}315 - 28{,}569 = ?$
20. List the factors of 20.

1.	2.	3.	4.
5.	6.	7.	8.
9.	10.	11.	12.
13.	14.	15.	16.
17.	18.	19.	20.

Lesson #36

1. What will be the time 80 minutes after noon?
2. $47{,}245 \div 5 = ?$
3. Put $\frac{2}{6}$ in simplest form.
4. $\frac{1}{3} + \frac{1}{5} = ?$
5. Trisha is 4 feet 8 inches tall. What is Trisha's height in inches?
6. How many degrees are in a straight angle?
7. Round 4,657 to the nearest hundred.
8. $13{,}506 \times 6 = ?$
9. Find the volume of the prism.

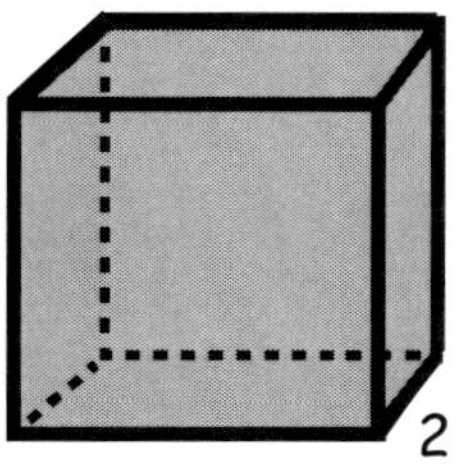

10. List the factors of 20.
11. Draw intersecting lines.
12. A(n) ____________ has six sides.
13. Two figures with the same size and shape are ___________.
14. Write $4\frac{1}{2}$ as an improper fraction.
15. On the Fahrenheit scale, water boils at ________.
16. $\frac{4}{7} \bigcirc \frac{5}{8}$
17. How many millimeters are in 3 meters?
18. $20{,}000 - 8{,}316 = ?$
19. $456 \div 24 = ?$
20. The prime factorization of a number is $2 \times 3^2 \times 5$. What is the number?

1.	2.	3.	4.
5.	6.	7.	8.
9.	10.	11.	12.
13.	14.	15.	16.
17.	18.	19.	20.

Lesson #37

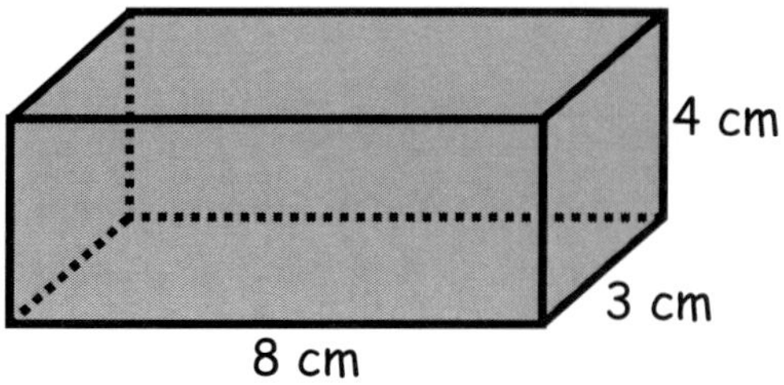

1. Find the volume of this rectangular solid.
2. Make a factor tree for 72.
3. Find the LCM of 9 and 14.
4. The Celsius freezing point of water is ________.
5. What is the probability of rolling a number greater than 2 on one roll of a die?
6. 345,119 + 762,344 = ?
7. Numbers with only two factors are called ____ numbers.
8. $\frac{5}{7} \bigcirc \frac{9}{10}$

9. Write *six and two tenths* as a decimal.
10. 806 − 279 = ?
11. How many feet are in 4 yards?
12. 374 × 22 = ?
13. Randy bought 12 comic books for his collection. Then he gave 3 comic books to his brother. Randy now has 38 comic books. How many did Randy have before he bought the 12?
14. List the factors of 16.
15. 6,020 ÷ 6 = ?
16. Put $\frac{8}{12}$ in simplest form.
17. Find the mean of 18, 25, 15, 29, and 18.
18. Write $\frac{14}{3}$ as a mixed number.
19. Half of the diameter of a circle is called the ________.
20. $\frac{3}{5} + \frac{1}{2} = ?$

1.	2.	3.	4.
5.	6.	7.	8.
9.	10.	11.	12.
13.	14.	15.	16.
17.	18.	19.	20.

Lesson #38

1. Draw perpendicular lines.
2. Find the area of the square.

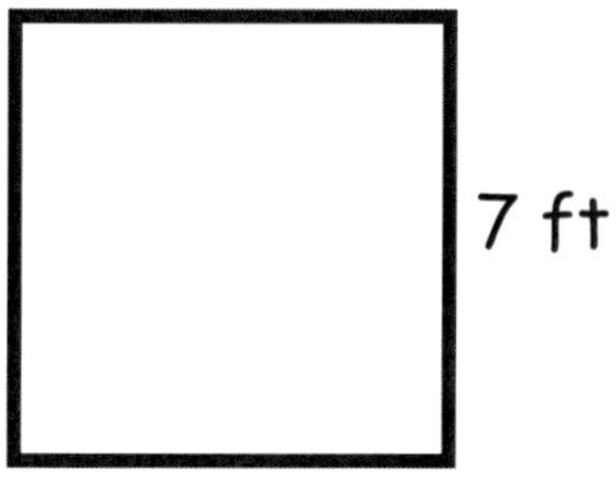

3. 60,000 − 38,515 = ?
4. Round 7,866,213 to the nearest hundred thousand.
5. Find the GCF of 12 and 18.
6. How many feet are in 2 miles?

7. 17.1 + 21.4 = ?
8. $\frac{3}{4} + \frac{1}{3} = ?$
9. Put $\frac{3}{24}$ in simplest form.
10. 748 ÷ 63 = ?
11. Write $6\frac{3}{7}$ as an improper fraction.
12. 13,642 × 3 = ?
13. Make a factor tree for 42.
14. Find the range of 31, 86, 92, 14, and 100.
15. Explain how to find the mean of a set of numbers.
16. How many centimeters are in 8 meters?
17. Which digit is in the ten millions place in 19,457,302?
18. Rachel has 2 pounds of sour cream. If she uses 14 ounces to make a coffee cake, how many ounces of sour cream does she have left?
19. $\frac{4}{5} \bigcirc \frac{6}{8}$
20. Find the volume of the figure.

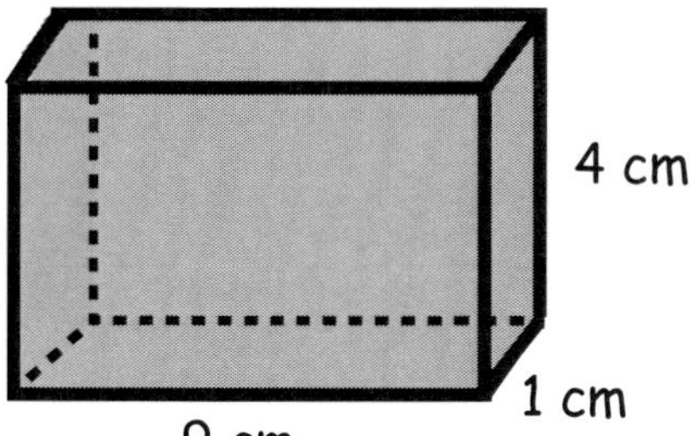

1.	2.	3.	4.
5.	6.	7.	8.
9.	10.	11.	12.
13.	14.	15.	16.
17.	18.	19.	20.

Lesson #39

1. Is this figure a polygon?

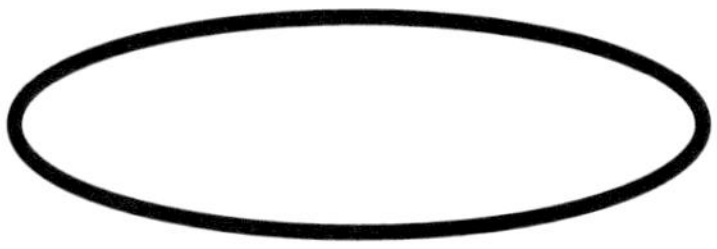

2. $\frac{3}{7} + \frac{1}{3} = ?$

3. Draw an obtuse angle.

4. $88 \times 34 = ?$

5. Write $\frac{33}{6}$ as a mixed number.

6. Find the perimeter of the quadrilateral.

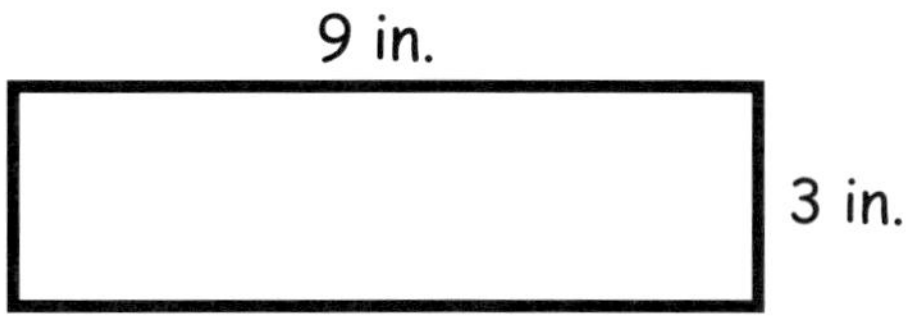

7. $97 \div 23 = ?$

8. Find the LCM of 6 and 15.

9. Is 56 a prime or a composite number?

10. How many millimeters are in 9 meters?

11. $331 + 469 = ?$

12. If the diameter of a circle is 22 m, what is the radius?

13. Grandma put a turkey in the oven at 11:30 a.m. She took it out of the oven at 3:45 p.m. How long did the turkey cook?

14. $\frac{4}{9} = \frac{?}{27}$

15. Write $5\frac{1}{7}$ as an improper fraction.

16. Find the mode of 37, 55, 18, 37, and 62.

17. List the factors of 20.

18. How many ounces are in 4 pounds?

19. Make a factor tree for 64.

20. The ________ is the answer in a subtraction problem.

1.	2.	3.	4.
5.	6.	7.	8.
9.	10.	11.	12.
13.	14.	15.	16.
17.	18.	19.	20.

Lesson #40

1. $\frac{3}{8}+\frac{1}{4}=?$

2. An angle that measures less than 90° is called ___________.

3. Write the time 7 minutes before 3:00.

4. Find the volume of the solid to the right.

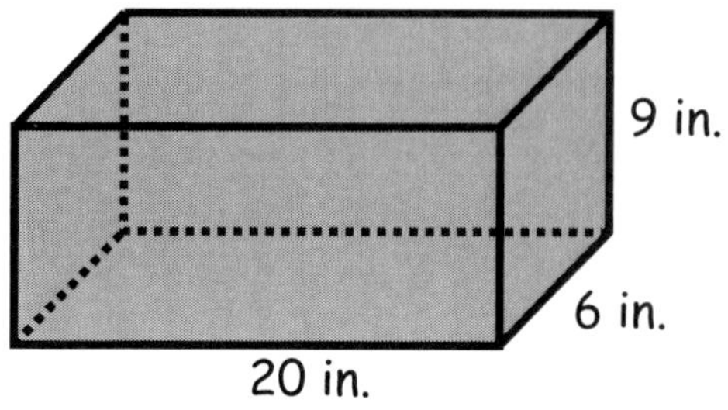

5. Round 38,204,516 to the nearest thousand.

6. $5{,}382 \div 26 = ?$

7. Write $\frac{15}{2}$ as a mixed number.

8. Draw a pentagon.

9. Find the median of 37, 48, 10, 19, and 41.

10. Write the odd numbers between 50 and 58.

11. Put $\frac{15}{20}$ in simplest form.

12. $800 \times 600 = ?$

13. How many quarts are in 4 gallons?

14. $\frac{9}{11} \bigcirc \frac{8}{9}$

15. $34{,}818 + 28{,}963 = ?$

16. Mark bought a television for $850. This price included $59 for tax and a discount of $124. What was the original price before the tax and the discount?

17. Draw parallel, horizontal lines.

18. Write 16.31 using words.

19. $801 - 667 = ?$

20. What is the probability that the spinner will land on an odd number?
A number less than 5?

1.	2.	3.	4.
5.	6.	7.	8.
9.	10.	11.	12.
13.	14.	15.	16.
17.	18.	19.	20.

Lesson #41

1. How many inches are in 3 yards?

2. Find the GCF of 12 and 18.

3. $51 + 18 + 7 = ?$

4. Round 5,304,298 to the nearest hundred.

5. Allie had 6 yards of rope. Holly had 144 inches of rope. How many more inches of rope did Allie have than Holly?

6. The answer in a multiplication problem is called the __________.

7. List the factors of 18.

8. Write $9\frac{2}{9}$ as an improper fraction.

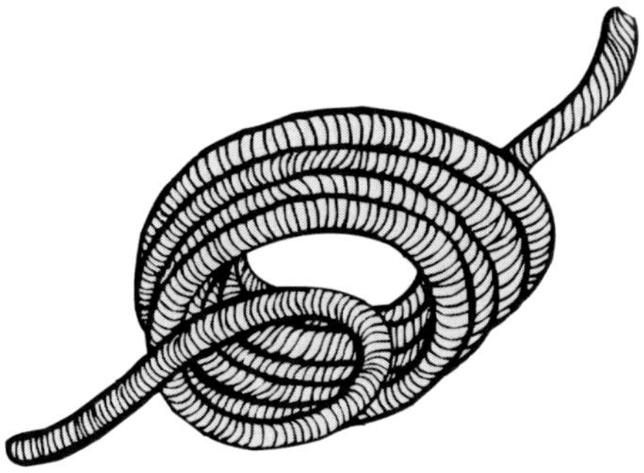

9. $43{,}215 \times 5 = ?$

10. Find the mean of 48, 72, 48, 83, and 49.

11. How many degrees are in a straight angle?

12. $30{,}000 - 19{,}471 = ?$

13. The boiling point of water, in degrees Celsius, is _________.

14. Find the area of a square whose sides each measure 8 inches.

15. The prime factorization for a number is $2^2 \times 3^2 \times 5$. What is the number?

16. $\frac{5}{6} + \frac{2}{3} = ?$

17. How many pounds are 6 tons?

18. Draw intersecting lines.

19. Sixty years are ________ decades.

20. $10{,}545 \div 5 = ?$

1.	2.	3.	4.
5.	6.	7.	8.
9.	10.	11.	12.
13.	14.	15.	16.
17.	18.	19.	20.

Lesson #42

1. $15 \times 6 = ?$
2. If it is 5:35 now, what time will it be in 2 hours and 25 minutes?
3. On the Fahrenheit scale, water freezes at __________.
4. $787{,}156 + 156{,}324 = ?$
5. $50 \div 8 = ?$
6. $6{,}318 - 4{,}756 = ?$
7. Write $6\frac{2}{5}$ as an improper fraction.
8. Draw a vertical line.
9. How many feet are in a mile?
10. $\frac{3}{5} \bigcirc \frac{4}{7}$
11. Put $\frac{7}{21}$ in simplest form.

12. Tina is 5 feet, 4 inches tall. What is Tina's height in inches?
13. List the factors of 16.
14. How many centuries are the same as 400 years?
15. Write *twenty-two and three tenths* as a decimal.
16. Make a factor tree for 30.
17. $5\frac{5}{7} + \frac{2}{3} = ?$
18. Find $\frac{1}{5}$ of 35.
19. Which city has the greatest difference between its normal high and low temperatures?
20. Which two cities have the most similar ranges in temperature?

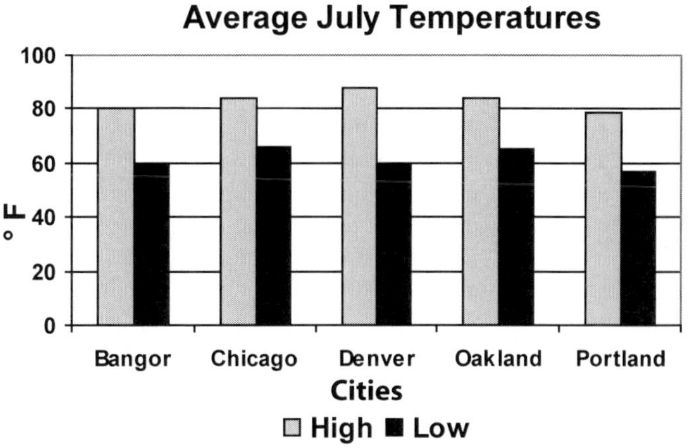

1.	2.	3.	4.
5.	6.	7.	8.
9.	10.	11.	12.
13.	14.	15.	16.
17.	18.	19.	20.

Lesson #43

1. Jessica used $\frac{1}{2}$ cup of sugar and $\frac{1}{8}$ cup of flour in a recipe. How many cups of ingredients did she use altogether?

2. 18 + ___ = 78

3. Find the volume of the figure.

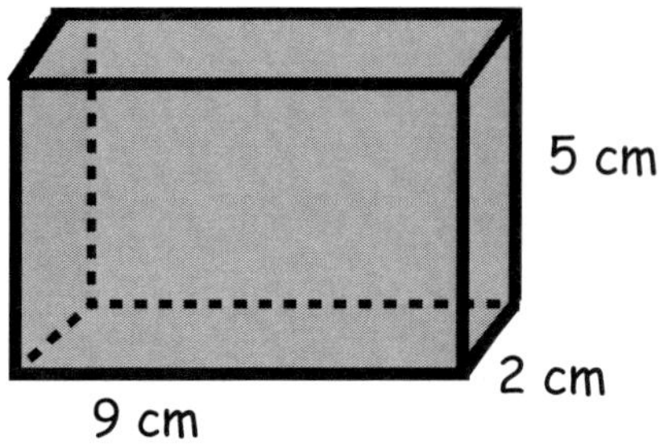

4. $57 \times 43 = ?$

5. Which is less, 50 feet or 12 yards?

6. $1\frac{4}{5} + 2\frac{1}{3} = ?$

7. Write the first four prime numbers.

8. $21{,}416 - 8{,}377 = ?$

9. Write $\frac{37}{4}$ as a mixed number.

10. The top number in a fraction is called the __________.

11. How many seconds are in 2 hours?

12. $45{,}065 \div 5 = ?$

13. Find $\frac{1}{3}$ of 21.

14. Round 5,203,776 to the nearest million.

15. Find the mean of 35, 31, 36, 34, and 34.

16. $\frac{7}{8} \bigcirc \frac{3}{4}$

17. Find the LCM of 8 and 12.

18. How many grams are in 8 kilograms?

19. List the factors of 24.

20. Write 12.07 using words.

1.	2.	3.	4.
5.	6.	7.	8.
9.	10.	11.	12.
13.	14.	15.	16.
17.	18.	19.	20.

Lesson #44

1. 80 − ___ = 25
2. Draw a ray.
3. Round 458,219 to the nearest ten thousand.
4. 837,222 + 768,354 = ?
5. Aunt Sarah used $2\frac{1}{4}$ yards of blue cloth and $1\frac{3}{8}$ yards of red cloth to make her niece a costume for school. How much cloth did Aunt Sarah use altogether?
6. Find the median of 11, 9, 22, 16, and 7.
7. 4,000 − 966 = ?
8. Nine centuries are _______ years.
9. Write the time 20 minutes before noon.
10. $6\frac{1}{4} + 8\frac{2}{5} = ?$
11. 33 × 29 = ?
12. Find the GCF of 12 and 14.
13. The answer to a division problem is the __________.
14. 15,813 ÷ 7 = ?
15. How many millimeters are in 5 meters?
16. 47.3 − 29.7 = ?
17. Find the mean of 43, 37, 43, 38, and 39.
18. Water freezes at _______ degrees Celsius.
19. List the factors of 12.
20. Write the formula for finding the area of a rectangle.

1.	2.	3.	4.
5.	6.	7.	8.
9.	10.	11.	12.
13.	14.	15.	16.
17.	18.	19.	20.

Lesson #45

1. $\frac{5}{8} \bigcirc \frac{9}{11}$
2. A six-sided polygon is called a(n) ______.
3. Write $3\frac{4}{7}$ as an improper fraction.
4. How many degrees are in a straight angle?

5. $4,325 \times 6 = ?$
6. Write 14.09 using words.
7. $205 - 78 = ?$
8. Make a factor tree for 32.
9. How many teaspoons are in 8 tablespoons?
10. Put $\frac{7}{21}$ in simplest form.
11. $\frac{9}{10} - \frac{3}{10} = ?$
12. Round 3,209,866 to the nearest hundred.
13. $621 + 588 + 93 = ?$
14. It is 1:15 now. What time was it 7 hours ago?
15. The probability of getting more than 2 on one roll of a die is _______.
16. Are these shapes similar or congruent?

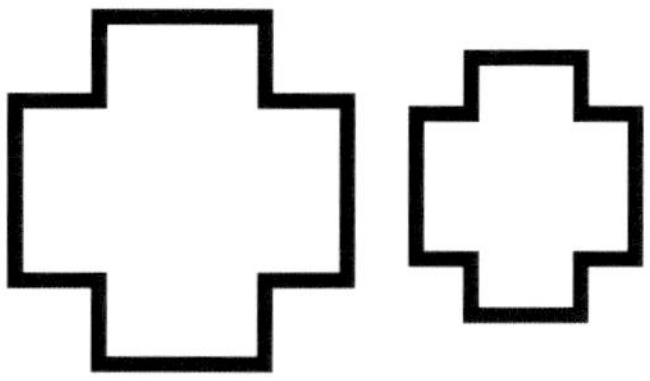

17. $5\frac{1}{6} + 3\frac{1}{2} = ?$
18. If the diameter of a circle is 18 mm, what is the radius?
19. There are 16 roses in the first row of our garden. The second row has 24 roses, the third row has 34 roses and the fourth row has 46 roses. How many roses are likely to be in the seventh row?
20. How many decades are 70 years?

1.	2.	3.	4.
5.	6.	7.	8.
9.	10.	11.	12.
13.	14.	15.	16.
17.	18.	19.	20.

Lesson #46

1. Find the mode of 56, 81, 13, 27, and 56.

2. Write the standard number for 50,000 + 4,000 + 500 + 60 + 9.

3. $8\frac{1}{9} + 2\frac{1}{3} = ?$

4. Find $\frac{1}{5}$ of 45.

5. Define *polygon.*

6. List the factors of 15.

7. How many feet are in 3 miles?

8. The answer to an addition problem is called the ________.

9. 34,816 + 79,065 = ?

10. Find the volume of the cube.

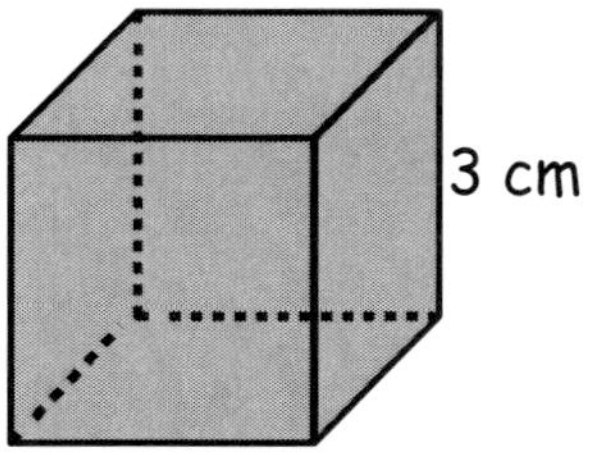

11. 6,000 − 1,586 = ?

12. Which has more sides, a hexagon or a pentagon? How many more?

13. What time will it be in 35 minutes, if it is 10:25?

14. How many pints are in 3 quarts?

15. 57 × 14 = ?

16. 836 ÷ 25 = ?

17. Find the GCF of 16 and 24.

Ticket Sales for County Fair

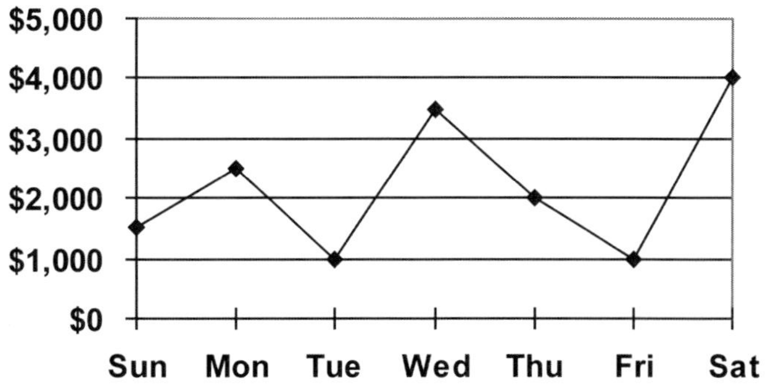

18. Draw perpendicular lines.

19. Between which two days did ticket sales show the greatest increase?

20. What were the total ticket sales for this week?

1.	2.	3.	4.
5.	6.	7.	8.
9.	10.	11.	12.
13.	14.	15.	16.
17.	18.	19.	20.

Lesson #47

1. Find the range of 87, 42, 67, 99, and 31.

2. $34 \times 29 = ?$

3. Write the formula for finding the volume of a solid.

4. List the factors of 16.

5. $625 \div 25 = ?$

6. How many quarts are in 10 gallons?

7. Draw intersecting lines.

8. $\frac{2}{3} \bigcirc \frac{4}{5}$

9. Write the time 45 minutes before 5:00.

10. Find $\frac{1}{3}$ of 27.

11. $18.9 - 9.4 = ?$

12. $10 - 4\frac{2}{3} = ?$

13. Which digit is in the hundred thousands place in 6,219,453?

14. Find the perimeter of the regular pentagon.

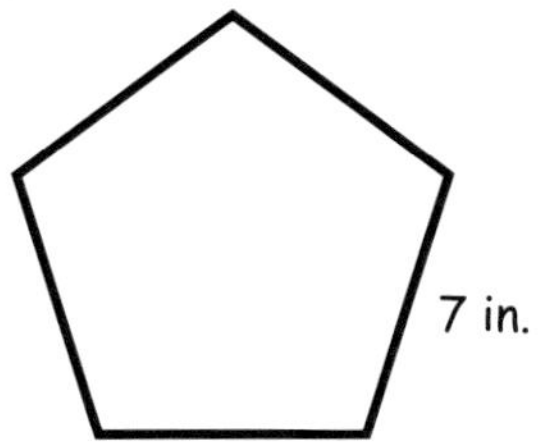

15. The boiling point of water is ____ degrees Celsius.

16. Find the LCM of 14 and 16.

17. Linda worked 6.5 hours on Wednesday and 8.25 hours on Friday. What is the total number of hours that she worked?

18. $5\frac{1}{5} + 6\frac{3}{4} = ?$

19. How many centimeters are 5 meters?

20. $375{,}816 + 48{,}932 = ?$

1.	2.	3.	4.
5.	6.	7.	8.
9.	10.	11.	12.
13.	14.	15.	16.
17.	18.	19.	20.

Lesson #48

1. Is 32 a prime or a composite number?

2. 80,000 − 39,176 = ?

3. Find the area of a square whose sides each measure 4 feet.

4. 480 × 24 = ?

5. How many cups are in 4 pints?

6. $\frac{3}{10} + \frac{2}{5} = ?$

7. Draw a line segment.

8. $\frac{3}{5} \bigcirc \frac{5}{6}$

9. 457 + 386 = ?

10. Round 4,877,231 to the nearest million.

11. Find $\frac{1}{8}$ of 56.

12. What time will it be in 1 hour and 45 minutes, if it is 5:45 now?

13. 148 ÷ 37 = ?

14. Put $\frac{5}{20}$ in simplest form.

15. Draw an obtuse angle.

16. If the diameter of a circle is 48 mm, what is the radius?

17. $13\frac{1}{9} - 10\frac{7}{9} = ?$

18. Figures with the same size and shape are ________.

19. How many feet are in 4 yards?

20. Eight dozen cookies will be divided evenly among 16 children. How many cookies will each child receive?

1.	2.	3.	4.
5.	6.	7.	8.
9.	10.	11.	12.
13.	14.	15.	16.
17.	18.	19.	20.

Lesson #49

1. Find the GCF of 14 and 21.
2. Draw a pentagon.
3. Find the area of the square.

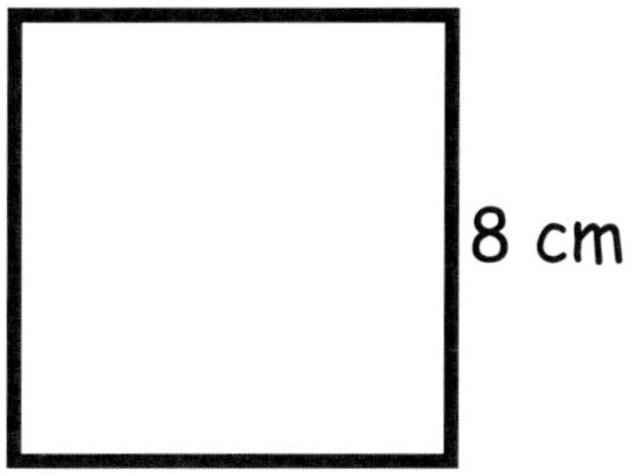

4. 35,776 + 48,213 = ?
5. $\frac{4}{5} \bigcirc \frac{7}{8}$
6. How many inches are in 4 feet?

7. $6\frac{1}{3} + 5\frac{2}{5} = ?$
8. 42,066 ÷ 6 = ?
9. Round 45,818,296 to the nearest ten thousand.
10. Find $\frac{1}{3}$ of 18.
11. Write *seven and two hundredths* as a decimal.
12. 5,071 − 3,698 = ?
13. Write $\frac{19}{5}$ as a mixed number.
14. Figures having the same shape, but different sizes are ________.
15. $8\frac{1}{9} - 6\frac{7}{9} = ?$
16. Put $\frac{15}{25}$ in simplest form.
17. 38,214 × 4 = ?
18. If Keith bought $\frac{3}{4}$ pound of candy and ate $\frac{1}{8}$ of it, how much is left?
19. List the factors of 24.
20. How many cups are in 4 pints?

1.	2.	3.	4.
5.	6.	7.	8.
9.	10.	11.	12.
13.	14.	15.	16.
17.	18.	19.	20.

Lesson #50

1. Define *composite number.*

2. How many degrees are in a straight angle?

3. Find the volume of this solid figure.

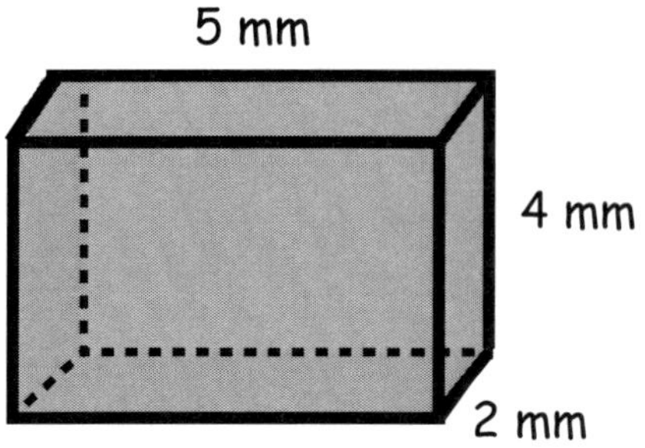

4. Find the GCF of 12 and 16.

5. $4 - 2\frac{3}{7} = ?$

6. On the Celsius scale, water boils at _________.

7. Make a factor tree for 60.

8. $474 + 398 + 512 = ?$

9. $10\frac{2}{3} + 8\frac{1}{4} = ?$

10. Find the range of 73, 24, 99, and 18.

11. How many millimeters are in 8 meters?

12. $5{,}436 \div 6 = ?$

13. $6.2 + 4.9 = ?$

14. Write $\frac{28}{6}$ as a mixed number.

15. $\frac{5}{9} \bigcirc \frac{7}{10}$

16. $70{,}000 - 28{,}453 = ?$

17. Put $\frac{9}{27}$ in simplest form.

18. A hexagon has ______ sides.

19. How many students were polled?

20. How many more students chose ice cream cone than sundae?

Favorite Ice Cream Treats Among Students	
Ice Cream Cone	245
Sundae	65
Banana Split	40
Milkshake	175
Ice Cream Float	85

1.	2.	3.	4.
5.	6.	7.	8.
9.	10.	11.	12.
13.	14.	15.	16.
17.	18.	19.	20.

Lesson #51

1. Susie is 4 feet 5 inches tall. Jessica is 51 inches tall. Which girl is taller? By how much?

2. Draw an obtuse angle.

3. 473,209 + 886,775 = ?

4. $16\frac{1}{9} - 12\frac{7}{9} = ?$

5. Write the next number in the sequence. 86, 81, 76, …

6. $350 \div 70 = ?$

7. $39 \times 64 = ?$

8. Which is larger, 800 centimeters or 6 meters?

9. How many minutes are in 4 hours?

10. The area of a square is 81 square inches. What is the length of each side?

11. 6,005 − 1,297 = ?

12. The answer to a division problem is the __________.

13. Find the LCM of 12 and 14.

14. $8\frac{1}{2} + 9\frac{2}{5} = ?$

15. Put $\frac{4}{12}$ in simplest form.

16. Round 7,213,886 to the nearest thousand.

17. How many years are 8 centuries?

18. $\frac{5}{12} \bigcirc \frac{7}{9}$

19. The distance across a circle, through the center, is the ________.

20. If it is 3:40 now, what time was it 4 hours and 10 minutes ago?

1.	2.	3.	4.
5.	6.	7.	8.
9.	10.	11.	12.
13.	14.	15.	16.
17.	18.	19.	20.

Lesson #52

1. Write the formula for finding the perimeter of a regular polygon.
2. $432 \times 25 = ?$
3. $5.23 + 8.7 = ?$
4. List the factors of 18.
5. Find the median and the mode of 24, 46, 32, 18, and 24.
6. $40{,}000 - 32{,}888 = ?$
7. $6\frac{1}{8} - 2\frac{5}{8} = ?$
8. Write $\frac{43}{8}$ as a mixed number.
9. How many feet are in 7 yards?
10. Make a factor tree for 42.
11. $8{,}624 \div 4 = ?$

12. Define *prime number.*
13. There are _____ degrees in a right angle.
14. $\frac{8}{9} \bigcirc \frac{3}{4}$
15. Write $4\frac{2}{7}$ as an improper fraction.
16. Is 19 a prime or a composite number?
17. Draw perpendicular lines.
18. Find $\frac{1}{5}$ of 50.
19. $4\frac{2}{3} + 6\frac{1}{5} = ?$
20. Our school is selling tickets to the Haunted Hayride. You can purchase 10 tickets for $1.00 or 2 tickets for $0.25. Which is the better buy? Explain.

1.	2.	3.	4.
5.	6.	7.	8.
9.	10.	11.	12.
13.	14.	15.	16.
17.	18.	19.	20.

Lesson #53

1. What will be the time nine minutes before 4:00?

2. $8 - 3\frac{2}{7} = ?$

3. Find the LCM of 9 and 15.

4. Put $\frac{3}{21}$ in simplest form.

5. On the Fahrenheit scale, the temperature of boiling water is _______.

6. Write *fifteen and twenty-one hundredths* as a decimal number.

7. Find the volume of the rectangular solid.

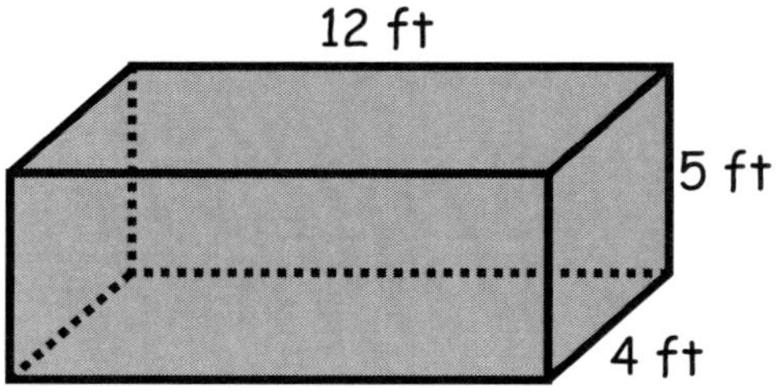

8. $4\frac{1}{7} + 8\frac{1}{3} = ?$

9. What is the probability of rolling a prime number on one roll of a die?

10. List the factors of 12.

11. Find $\frac{1}{6}$ of 36.

12. Draw intersecting lines.

13. Write the formula for finding the area of a rectangle.

14. Which digit is in the millions place in 36,021,547?

15. How many grams are in 6 kilograms?

16. Find the range of 38, 100, 19, and 45.

17. An eight-sided polygon is a(n) ________.

18. 503 − 199 = ?

19. 49 × 25 = ?

20. Harris Middle School has 4 classes at each grade level, sixth through eighth grade. If there is an average of 26 students in each classroom, about how many students attend Harris Middle School?

1.	2.	3.	4.
5.	6.	7.	8.
9.	10.	11.	12.
13.	14.	15.	16.
17.	18.	19.	20.

Lesson #54

1. Draw a ray.
2. $364 + 287 = ?$
3. Find the perimeter of a square whose sides measure 3 feet each.
4. How many months are in 3 years?
5. Round 3,263,551 to the nearest hundred.
6. What do you call a number with only two factors?
7. $18 - 13\frac{5}{9} = ?$

8. $3{,}210 - 1{,}776 = ?$
9. Put $\frac{3}{15}$ in simplest form.
10. What will be the time 50 minutes after 6:00?
11. Is 42 a prime or a composite number?
12. $5\frac{1}{9} + 3\frac{1}{2} = ?$
13. Find the volume of this solid figure.

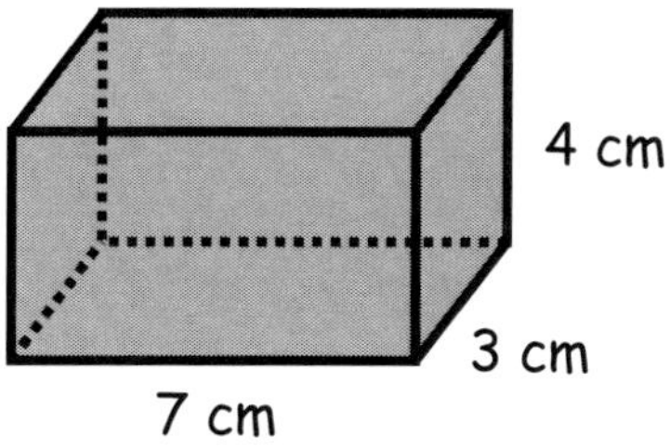

14. How many degrees are in a straight angle?
15. $87 \times 45 = ?$
16. $\frac{4}{11} \bigcirc \frac{5}{6}$
17. How many pounds are in 6 tons?
18. Make a factor tree for 90.
19. Find the median of 30, 19, 22, 16, and 41.
20. The Yummy Bear Candy Company sells three times as many chocolate covered pretzels as suckers. They sell twice as many gumballs as suckers. If they sell 3,750 chocolate covered pretzels, how many gumballs do they sell?

1.	2.	3.	4.
5.	6.	7.	8.
9.	10.	11.	12.
13.	14.	15.	16.
17.	18.	19.	20.

Lesson #55

1. Write 6.08 using words.
2. Find the perimeter of a pentagon if each side measures 8 inches.
3. $14\frac{2}{3} + 9\frac{1}{5} = ?$
4. Round 38,215,077 to the nearest ten thousand.
5. If the diameter of a circle is 12 mm, what is the radius?
6. Find the GCF of 10 and 30.
7. Water freezes at what Celsius temperature?
8. $235 \times 36 = ?$
9. $30{,}000 - 19{,}375 = ?$
10. $12\frac{1}{7} - 10\frac{5}{7} = ?$
11. How many feet are in 3 miles?
12. Draw a line of symmetry on the arrow.

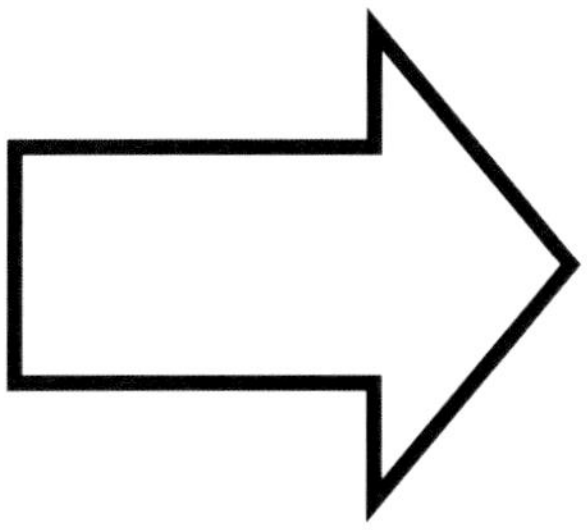

13. Write 44,235 in expanded form.
14. $840 \div 70 = ?$
15. Mr. Franco is harvesting his garden. The rectangular garden is thirty five feet by eighteen feet. What is the area of his garden?
16. Draw an acute angle.
17. Write $\frac{41}{5}$ as a mixed number.
18. How many centimeters are in 4 meters?

19. $337{,}218 + 887{,}993 = ?$
20. List the factors of 12.

1.	2.	3.	4.
5.	6.	7.	8.
9.	10.	11.	12.
13.	14.	15.	16.
17.	18.	19.	20.

Lesson #56

1. $73 + 26 + 17 = ?$
2. Water freezes at ____ degrees Fahrenheit.
3. $13 - 8\frac{5}{6} = ?$
4. $388{,}754 + 662{,}581 = ?$
5. Round 739,210,568 to the nearest hundred million.
6. Find the area of the square.

7. $7{,}230 - 3{,}986 = ?$
8. Two figures with the same size and shape are __________.
9. The middle number in a set of numbers is the __________.
10. $\frac{4}{9} \bigcirc \frac{3}{7}$
11. How many ounces are in 3 pounds?
12. Draw two similar squares.
13. $350 \div 50 = ?$
14. Make a factor tree for 40.
15. Write $3\frac{2}{9}$ as an improper fraction.
16. $38 \times 23 = ?$

17. Define *mode*.
18. What is the probability of rolling a number greater than four on one roll of a die?
19. Tonya, Kelly, and Sandy went shopping on Saturday. Tonya spent \$19.75. Kelly spent \$4.25 more than Sandy. Sandy spent \$1.50 less than Tonya. How much did they spend in all?
20. How many teaspoons are in 3 tablespoons?

1.	2.	3.	4.
5.	6.	7.	8.
9.	10.	11.	12.
13.	14.	15.	16.
17.	18.	19.	20.

Lesson #57

1. Write *nineteen and four tenths* as a decimal.

2. Draw parallel horizontal lines.

3. 53,218 + 69,774 = ?

4. $18\frac{1}{7} - 11\frac{6}{7} = ?$

5. Put $\frac{7}{49}$ in simplest form.

6. $49 \times 51 = ?$

7. How many pints are in 5 quarts?

8. The answer to a division problem is called the _________.

9. $\frac{8}{12} \bigcirc \frac{9}{10}$

10. Round 42,816,755 to the nearest thousand.

11. $5\frac{1}{3} + 9\frac{1}{4} = ?$

12. Find the volume of the rectangular prism.

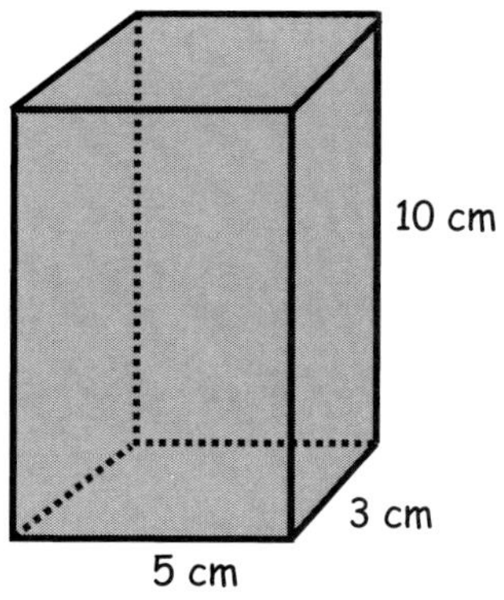

13. 600 − 299 = ?

14. The custodians have set up chairs for 27 third graders to watch a movie. There are 2 rows of six chairs, 3 rows of four chairs, and 3 rows of three chairs. Are there enough chairs for the 3rd graders?

15. Find the range of 86, 73, 28, 42, and 105.

16. 4,550 ÷ 50 = ?

17. How many tons are 6,000 pounds?

18. A polygon with four sides is called a(n) __________.

19. Write $7\frac{2}{3}$ as an improper fraction.

20. Find the LCM of 14 and 18.

1.	2.	3.	4.
5.	6.	7.	8.
9.	10.	11.	12.
13.	14.	15.	16.
17.	18.	19.	20.

Lesson #58

1. How many feet are in 5 miles?
2. $\frac{5}{8} \times \frac{16}{25} = ?$
3. Draw a right angle.
4. Find the perimeter of this rectangle.

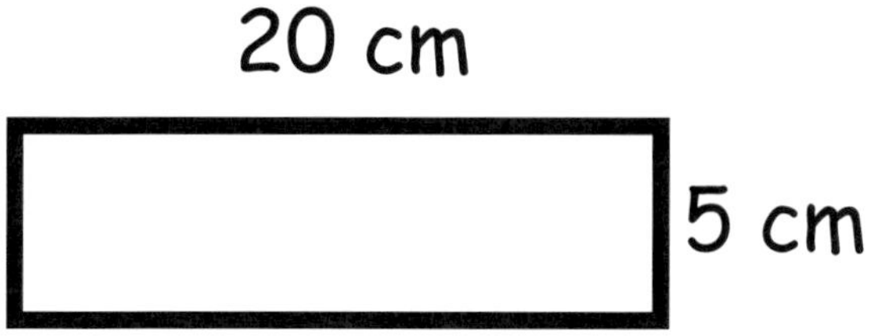

5. $17\frac{1}{8} - 9\frac{5}{8} = ?$
6. Write the decimal 12.3 using words.
7. 66,374 + 98,355 = ?
8. Round 2,455,919 to the nearest ten thousand.
9. $6\frac{2}{5} + 8\frac{1}{4} = ?$
10. 720 ÷ 30 = ?
11. Michael's tent is 5 feet high, George's tent is 2 yards high, and William's tent is 68 inches high. Whose tent is the highest?

12. 70,000 − 43,552 = ?
13. Make a factor tree for 32.
14. 331 × 28 = ?
15. Find the GCF of 12 and 18.
16. List the factors of 20.
17. 2,475 ÷ 5 = ?
18. If the radius of a circle is 12 mm, what is the diameter?
19. How many quarts are in 7 gallons?
20. $\frac{3}{8}$ ◯ $\frac{5}{7}$

1.	2.	3.	4.
5.	6.	7.	8.
9.	10.	11.	12.
13.	14.	15.	16.
17.	18.	19.	20.

Lesson #59

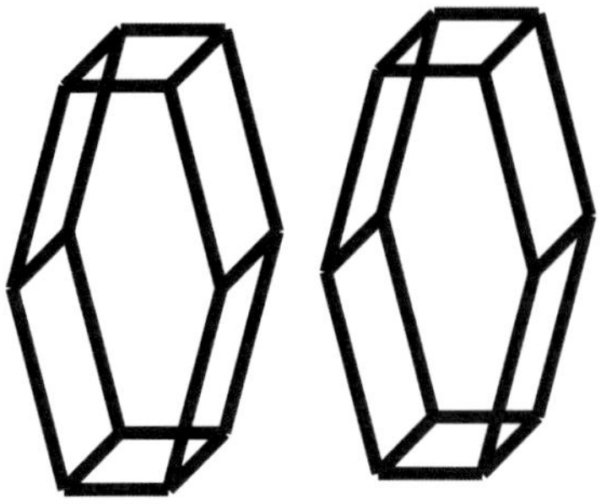

1. Are these shapes congruent?
2. 403,575 + 186,691 = ?
3. $5\frac{1}{7} + 3\frac{1}{3} = ?$
4. Put $\frac{10}{15}$ in simplest form.
5. A pack of 12 pencils costs $1.44. What is the cost of each pencil?
6. How many inches are in 2 yards?

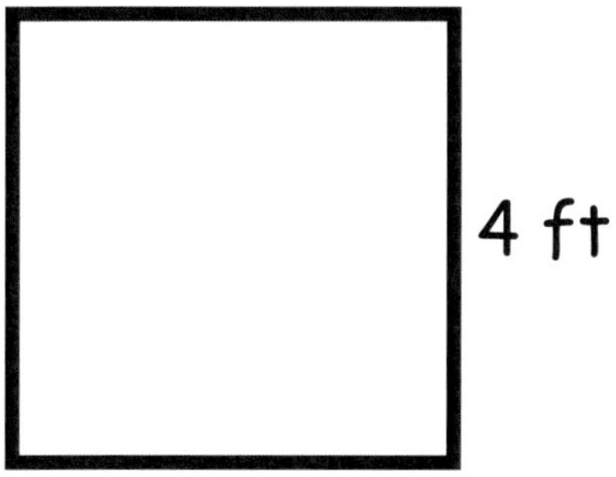

7. 403 − 196 = ?
8. Find the perimeter of this square.
9. $\frac{5}{7} = \frac{?}{21}$
10. 318 × 25 = ?
11. Round 93,114,587 to the nearest ten thousand.
12. A play began at 3:15 p.m. and lasted for 2 hours and 15 minutes. There was a 10 minute intermission. At what time did the play end?

13. Find the LCM of 8 and 12.
14. 87,256 ÷ 6 = ?
15. Write $\frac{53}{9}$ as a mixed number.
16. How many pints are in 4 quarts?
17. $15 - 10\frac{3}{5} = ?$
18. Find $\frac{1}{9}$ of 36.
19. Draw an obtuse angle.
20. Find the mean of 848, 327, and 838.

1.	2.	3.	4.
5.	6.	7.	8.
9.	10.	11.	12.
13.	14.	15.	16.
17.	18.	19.	20.

Lesson #60

1. Make a factor tree for 100.
2. 50,000 − 24,315 = ?
3. How many decades are 80 years?
4. $4\frac{2}{7} + 4\frac{1}{5} = ?$
5. Find the area of the quadrilateral.

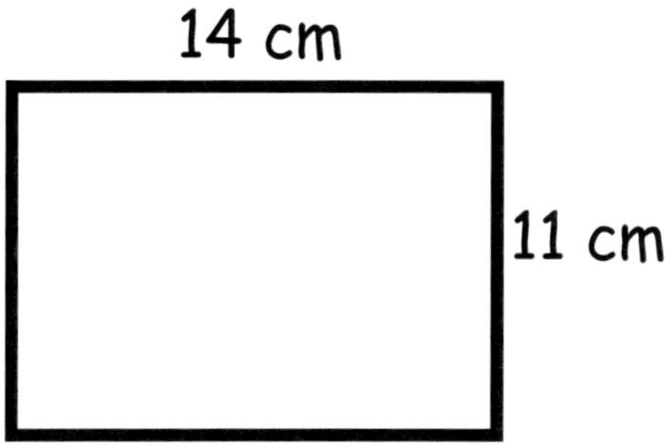

6. 3,600 ÷ 40 = ?
7. Write $3\frac{4}{7}$ as an improper fraction.
8. $\frac{5}{7} = \frac{35}{?}$
9. $16\frac{1}{10} - 10\frac{7}{10} = ?$
10. 14,819 × 9 = ?
11. Find the range of 75, 80, 100, 95, and 82.
12. Is $\frac{3}{7}$ in simplest form?
13. $\frac{4}{9} \bigcirc \frac{3}{8}$
14. Maureen charges $4 to walk a small dog and $6 to walk a large dog. If she walks 12 small dogs and 4 large dogs, what will she earn?
15. 45,712 + 22,949 = ?
16. A four-sided polygon is called a(n) __________.
17. Define *mode*.
18. $\frac{4}{5} + \frac{1}{5} = ?$
19. How many centimeters are 6 meters?
20. In a bag of seven marbles, 4 are red and 3 are green. What is the probability of picking a green one? A red one? A blue one?

1.	2.	3.	4.
5.	6.	7.	8.
9.	10.	11.	12.
13.	14.	15.	16.
17.	18.	19.	20.

Lesson #61

1. $8\frac{1}{5} + 3\frac{3}{4} = ?$

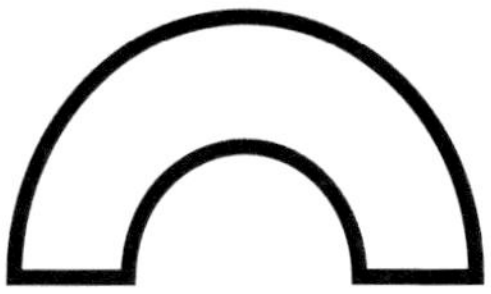

2. Is this figure a polygon?

3. $3{,}007 - 1{,}689 = ?$

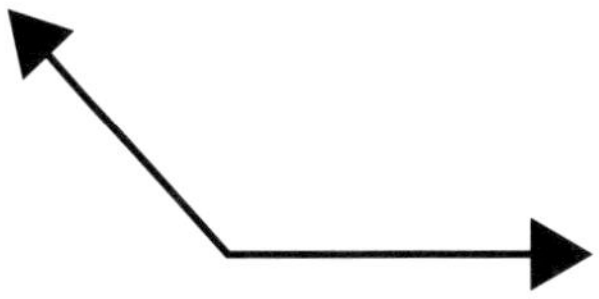

4. Identify this angle by type.

5. $\frac{5}{9} \times \frac{18}{20} = ?$

6. Round 364,531 to the nearest thousand.

7. Write $1\frac{5}{8}$ as an improper fraction.

8. Find the mean of 110, 117, 130, and 139.

9. Put $\frac{12}{18}$ in simplest form.

10. $4{,}863 \times 7 = ?$

11. Rick read half the amount of time that Rita read. Rita read one-third of the time that Larry read. Larry read twice as long as Tina. If Tina read for 3 hours, how long did Rick read?

12. Find the area of a square whose sides each measure 5 inches.

13. $897 \div 38 = ?$

14. How many cups are in 4 pints?

15. $\frac{3}{7} \bigcirc \frac{2}{5}$

16. Find the GCF of 14 and 21.

17. $35.7 + 8.92 = ?$

18. The middle number in a group of numbers is the _______.

19. $437{,}806 + 831{,}215 = ?$

20. Is the sum in item 19 an even or an odd number?

1.	2.	3.	4.
5.	6.	7.	8.
9.	10.	11.	12.
13.	14.	15.	16.
17.	18.	19.	20.

Lesson #62

1. Find the range of 37, 52, 97, and 15.
2. Write $\frac{45}{7}$ as a mixed number.
3. $35 \times 49 = ?$
4. $4\frac{1}{6} - 2\frac{5}{6} = ?$
5. $8{,}643 \div 3 = ?$
6. Find the area of the rectangle.

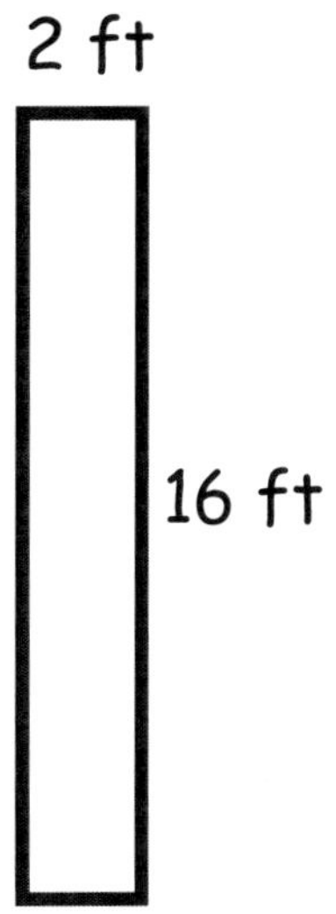

7. How many days are in 5 weeks?
8. Find the LCM of 10 and 12.
9. $80{,}000 - 43{,}292 = ?$
10. Which digit is in the ten millions place in 45,032,687?
11. Six pints are how many cups?
12. Define *median*.
13. $\frac{4}{9} = \frac{16}{?}$
14. $6\frac{1}{9} - 8\frac{1}{2} = ?$
15. Find $\frac{1}{7}$ of 56.

16. $37{,}414 + 28{,}065 = ?$
17. The answer to a multiplication problem is the __________.
18. What year was it, exactly 4 centuries before 1965?
19. The prime factorization of a number is $2^3 \times 3^2$. What's the number?
20. The Smith family wants to fence their yard. The yard is 96 meters long and 52 meters wide. How much fencing will they need?

1.	2.	3.	4.
5.	6.	7.	8.
9.	10.	11.	12.
13.	14.	15.	16.
17.	18.	19.	20.

Lesson #63

1. Write $6\frac{3}{5}$ as an improper fraction.
2. A pentagon has _____ sides.
3. $307 - 188 = ?$
4. $14\frac{2}{7} - 10\frac{6}{7} = ?$
5. Write the formula for finding the area of a rectangle.
6. If it is 5:15 now, what time will it be in 8 hours and 10 minutes?
7. $324 \times 23 = ?$
8. The answer to a subtraction problem is called the _______.
9. Draw intersecting lines.
10. Who read the most books over the 3 week period?
11. Which two students read the same number of books?

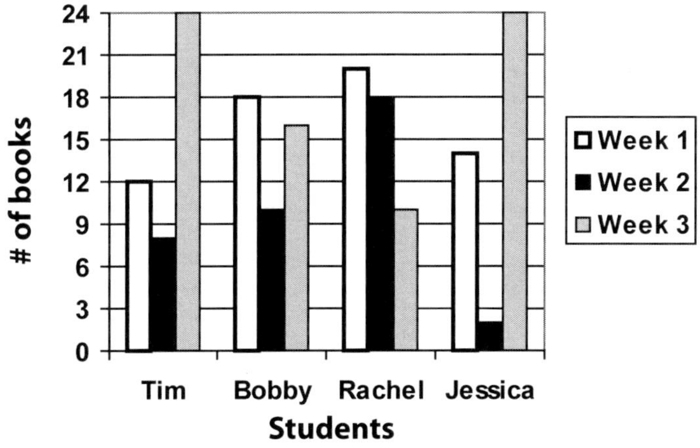

12. Put $\frac{8}{12}$ in simplest form.
13. $9\frac{2}{7} + 3\frac{1}{4} = ?$
14. List the factors of 10.
15. $\frac{4}{5} \times \frac{10}{16} = ?$
16. Find the volume of the cube.

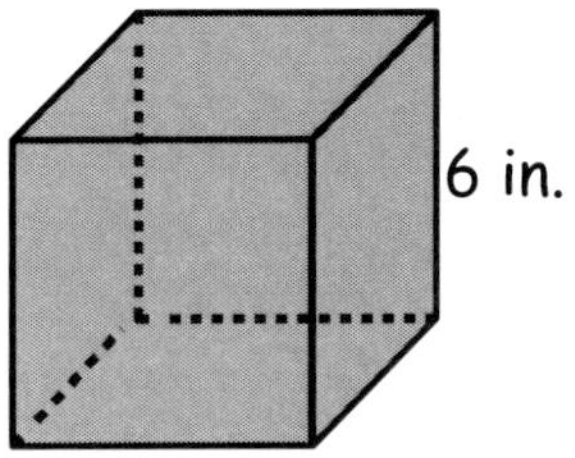

17. Which is greater, 3 miles or 16,000 feet? What is the difference between the two measurements?
18. $2,816 + 14,299 = ?$
19. Make a factor tree for 81.
20. Find the median of 8, 24, 16, 12, and 19.

1.	2.	3.	4.
5.	6.	7.	8.
9.	10.	11.	12.
13.	14.	15.	16.
17.	18.	19.	20.

Lesson #64

1. Find $\frac{1}{5}$ of 25.
2. How many yards are in a mile?
3. Draw a right angle.
4. Michael is 6 feet 5 inches tall. His brother is 73 inches tall. Who is taller? What is the difference in their heights?
5. $\frac{4}{9} - \frac{2}{9} = ?$
6. The boiling temperature of water, in degrees Fahrenheit, is ________.
7. $\frac{4}{7} = \frac{?}{21}$
8. Write 16.3 using words.
9. $\frac{7}{8} \bigcirc \frac{9}{10}$
10. $12 - 6\frac{4}{5} = ?$
11. Numbers that have only two factors are called ________ numbers.
12. What is the formula for finding the perimeter of a polygon?
13. $16{,}346 \div 6 = ?$
14. $263 \times 5 = ?$
15. $10\frac{1}{5} - 3\frac{3}{4} = ?$
16. How many feet are in 4 yards?
17. $6{,}031 - 3{,}446 = ?$
18. $3.7 + 21.9 = ?$

19. A closed figure made up of line segments is a(n) ________.
20. Make a factor tree for 80.

1.	2.	3.	4.
5.	6.	7.	8.
9.	10.	11.	12.
13.	14.	15.	16.
17.	18.	19.	20.

Lesson #65

1. $13 - 8\frac{3}{8} = ?$
2. How many feet are in 6 yards?
3. $\frac{5}{11} \div \frac{15}{22} = ?$
4. Put $\frac{12}{16}$ in simplest form.
5. $\frac{3}{8} \times \frac{10}{12} = ?$
6. 361,217 + 487,354 = ?
7. Write $3\frac{5}{7}$ as an improper fraction.
8. 334 × 25 = ?
9. Four books sit on a bookcase. The widths of the books are 1.3 cm, 0.6 cm, 2.1 cm, and 1.5 cm. What is the total width of these books?
10. $11\frac{2}{5} + 13\frac{1}{4} = ?$
11. Find the GCF of 12 and 30.
12. Draw a line segment.
13. 821 − 375 = ?
14. How many grams are in 7 kilograms?
15. $\frac{4}{9} \bigcirc \frac{2}{3}$
16. Find the median of 53, 16, 29, 31, and 46.
17. Write *six and four tenths* as a decimal.
18. Make a factor tree for 60.
19. 480 ÷ 60 = ?
20. Round 54,807,334 to the nearest million.

1.	2.	3.	4.
5.	6.	7.	8.
9.	10.	11.	12.
13.	14.	15.	16.
17.	18.	19.	20.

Lesson #66

1. $32{,}406 \times 6 = ?$

2. How many gallons are 8 quarts?

3. $3\frac{2}{9} + 5\frac{1}{2} = ?$

4. It is 3:40. What time will it be in 6 hours and 20 minutes?

5. $\$63.45 - \$18.99 = ?$

6. Draw an acute angle.

7. Find the volume of the rectangular prism.

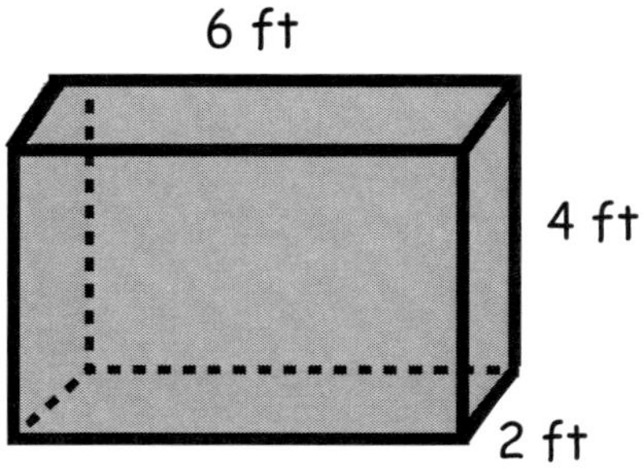

8. Put $\frac{14}{21}$ in simplest form.

9. Tyrone has 24 fish in his aquarium. Milo has one-fourth the number of fish that Tyrone has. Jerome has twice as many fish as Milo. How many fish do they have altogether?

10. $4{,}763 \div 3 = ?$

11. The answer to a subtraction problem is the _______.

12. $\frac{4}{7} \times \frac{14}{24} = ?$

13. $43.92 + 18.8 = ?$

14. Which digit is in the ten millions place in 17,206,345?

15. Find $\frac{1}{8}$ of 64.

16. Find the mode of 18, 23, 46, 29, and 23.

17. $\frac{4}{7} \bigcirc \frac{5}{9}$

18. List the factors of 21.

19. How many feet are in 2 miles?

20. $18\frac{1}{6} - 12\frac{5}{6} = ?$

1.	2.	3.	4.
5.	6.	7.	8.
9.	10.	11.	12.
13.	14.	15.	16.
17.	18.	19.	20.

Lesson #67

1. What type of angle is shown?

2. How many minutes are in 5 hours?

3. Write $\frac{32}{6}$ as a mixed number.

4. 3,415,291 + 8,229,804 = ?

5. Marshall has two dogs. One of the dogs weighs twice as much as the other one. If the total weight of the two dogs is 60 pounds, how much does each dog weigh?

6. $2\frac{2}{7} + 3\frac{1}{4} = ?$

7. $\frac{5}{8} \div \frac{5}{6} = ?$

8. What is water's boiling point on the Celsius scale?

9. Draw intersecting lines.

10. Find the perimeter of the regular pentagon.

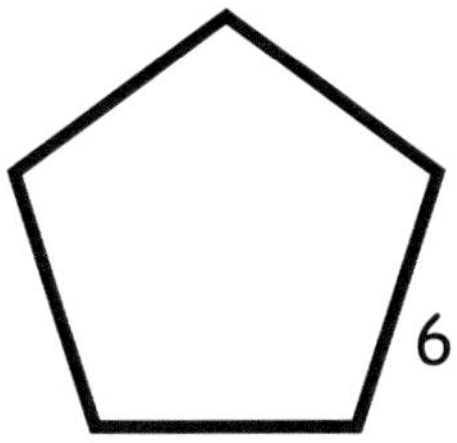

11. 70,000 − 27,316 = ?

12. $77 \times 42 = ?$

13. $\frac{4}{9} = \frac{?}{27}$

14. What time was it 6 hours ago, if it is 2:00 now?

15. Give the name of the figure to the right.

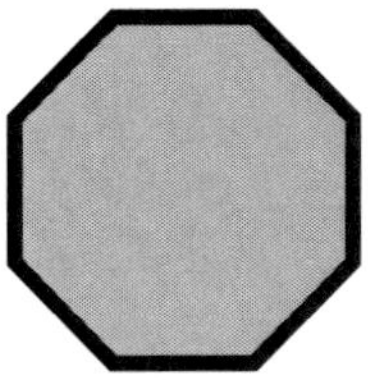

16. How many pounds are 96 ounces?

17. Find the radius of a circle whose diameter is 24 mm.

18. Find the LCM of 12 and 14.

19. $45{,}235 \div 5 = ?$

20. Which is heavier 300 grams or 300 kilograms?

1.	2.	3.	4.
5.	6.	7.	8.
9.	10.	11.	12.
13.	14.	15.	16.
17.	18.	19.	20.

Lesson #68

1. Put $\frac{12}{16}$ in simplest form.
2. $302 - 176 = ?$
3. Find the volume of a cube whose sides measure 6 inches.
4. Define *prime number.*
5. $11\frac{1}{2} - 6\frac{4}{5} = ?$
6. $86 \times 34 = ?$
7. Make a factor tree for 24.
8. List the factors of 30.
9. There are _____ degrees in a straight angle.
10. $361,247 + 285,932 = ?$
11. Write $9\frac{2}{3}$ as an improper fraction.
12. Draw a line.
13. Four centuries are __________ years.
14. Find the mean of 23, 54, 87, and 84.
15. Round 86,345,219 to the nearest ten million.
16. Numbers with only two factors are ___________ numbers.
17. $280 \div 41 = ?$
18. Find $\frac{1}{5}$ of 35.
19. $8.05 - 0.46 = ?$
20. How many students prefer football over gymnastics?

    Baseball is the favorite sport of how many?

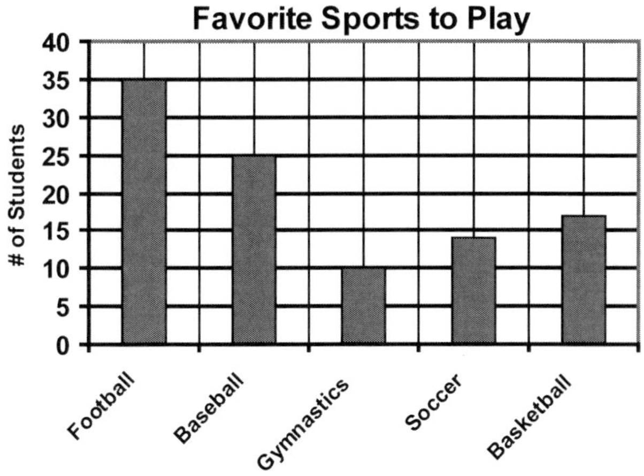

1.	2.	3.	4.
5.	6.	7.	8.
9.	10.	11.	12.
13.	14.	15.	16.
17.	18.	19.	20.

Lesson #69

1. 568,904 + 35,688 = ?

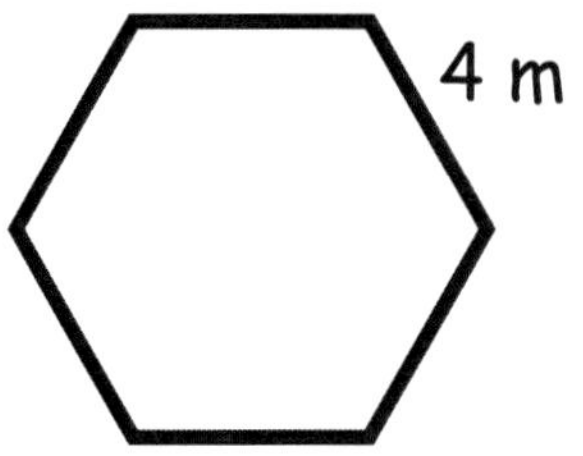

2. Draw two similar rectangles.

3. 245 × 31 = ?

4. Find the perimeter of the hexagon.

5. Find the GCF and the LCM of 12 and 18.

6. If it is 4:20 now, what time will it be in 5 hours and 15 minutes?

7. $\frac{5}{9} \bigcirc \frac{7}{8}$

8. How many inches are in a yard?

9. $\frac{6}{7} \times \frac{14}{24} = ?$

10. 7,023 − 4,889 = ?

11. 481 ÷ 14 = ?

12. Find the range and the median of 34, 88, 92, 13, and 56.

13. How many quarts are in 6 gallons?

14. Ronald wants to buy a new watch that costs $195. He has $75. If he saves $30 each week, in how many weeks will he be able to buy the watch?

15. Round 526,772,098 to the nearest hundred million.

16. $24 - 19\frac{7}{8} = ?$

17. Draw an obtuse angle.

18. Put these decimals in order from least to greatest. 2.3 2.03 2.43

19. What is the probability of rolling a prime number on a single roll of a die?

20. Is 42 a prime or a composite number?

1.	2.	3.	4.
5.	6.	7.	8.
9.	10.	11.	12.
13.	14.	15.	16.
17.	18.	19.	20.

Lesson #70

1. Put $\frac{12}{18}$ in simplest form.
2. $45{,}877 \times 3 = ?$
3. How many pounds are 8 tons?
4. Write *seventy and three hundredths* as a decimal.
5. $50{,}000 - 29{,}444 = ?$
6. Which digit is in the ten thousands place in 1,679,032?
7. What will be the time 35 minutes after noon?
8. Jason is 6 feet 7 inches tall. What is Jason's height in inches?
9. What is the area of the rectangle?

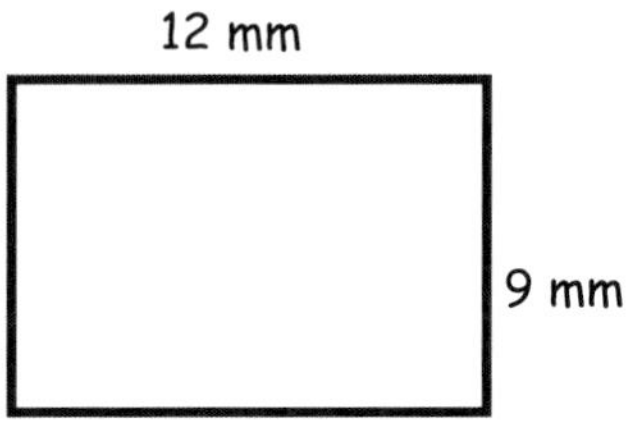

10. Draw perpendicular lines.
11. $35.6 - 19.37 = ?$
12. $8\frac{1}{3} + 5\frac{3}{5} = ?$
13. How many degrees are in a right angle?
14. $678 + 984 + 139 = ?$
15. $45{,}675 \div 5 = ?$
16. How many nickels are in $4?
17. $\frac{5}{9} \times \frac{12}{15} = ?$

18. The answer to an addition problem is known as the __________.
19. At the local pet store there are twice as many dogs as cats. There are twice as many cats as there are birds. If there are 16 dogs, how many birds are in the pet store?
20. Find $\frac{3}{5}$ of 35.

1.	2.	3.	4.
5.	6.	7.	8.
9.	10.	11.	12.
13.	14.	15.	16.
17.	18.	19.	20.

Lesson #71

1. $67 \times 24 = ?$

2. How many cups are in 7 pints?

3. Define *congruent.*

4. Classify the angle by type.

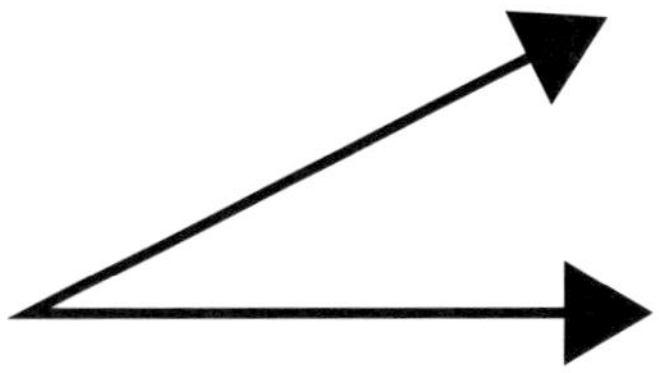

5. $12\frac{1}{6} - 10\frac{5}{6} = ?$

6. The top number in a fraction is the _________.

7. From January 1st to June 1st is how many months?

8. Put $\frac{10}{12}$ in simplest form.

9. $314 - 188 = ?$

10. $1\frac{1}{2} \times 2\frac{2}{3} = ?$

11. $\frac{5}{6} \div \frac{5}{3} = ?$

12. Karen's ballet class begins at 4:00 p.m. She needs 10 min. to walk to the bus stop, 21 min. for the ride and 4 min. to walk to class. What time should she leave home so she gets to class on time?

13. $478{,}903 + 551{,}266 = ?$

14. $\frac{5}{6} = \frac{25}{?}$

15. $47 + \_\_\_ = 96$

16. How many millimeters are in 9 meters?

17. Draw a ray.

18. $684 \div 24 = ?$

19. Write 12.46 using words.

20. Order these decimals from greatest to least. 7.15 7.015 7.51 7.051

1.	2.	3.	4.
5.	6.	7.	8.
9.	10.	11.	12.
13.	14.	15.	16.
17.	18.	19.	20.

Lesson #72

1. Write $\frac{23}{7}$ as a mixed number.

2. $705 \div 25 = ?$

3. Two figures having the same shape, but different sizes are _________.

4. $515 - 266 = ?$

5. How many teaspoons are in 9 tablespoons?

6. $89 + 53 + 77 = ?$

7. On the Fahrenheit temperature scale, water freezes at ________.

8. What time was it 4 hours and 5 minutes ago, if it is 1:15 now?

9. Find the LCM of 9 and 15.

10. $\frac{7}{9} \bigcirc \frac{9}{10}$

11. Find the area of the square.

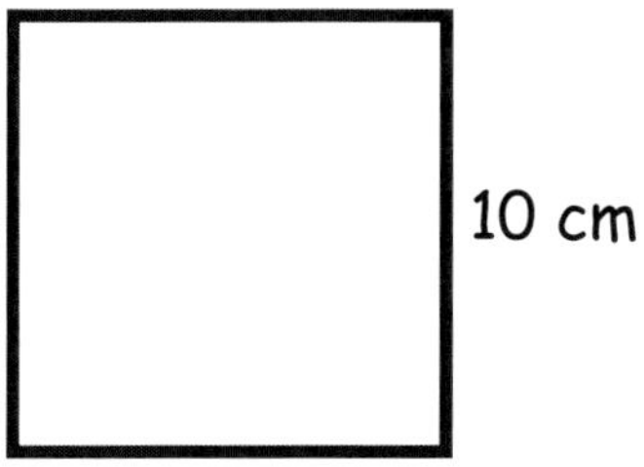

12. $5\frac{2}{5} + 6\frac{2}{3} = ?$

13. $9.4 - 6.88 = ?$

14. Find the mode of 44, 67, 29, 91, and 67.

15. Make a factor tree for 40.

16. $93 \times 35 = ?$

17. List the factors of 18.

18. Book Basement is selling every book for half price. Fran wants to buy four books that were originally priced at \$12.50, \$16, \$14.30, and \$10.44. What will be Fran's total?

19. Which digit is in the thousands place in 65,348,912?

20. Sixty years are ______ decades.

1.	2.	3.	4.
5.	6.	7.	8.
9.	10.	11.	12.
13.	14.	15.	16.
17.	18.	19.	20.

Lesson #73

1. $832 + 977 = ?$

2. Find the LCM of 12 and 18.

3. Put $\frac{18}{24}$ in simplest form.

4. Which factors of 14 are also factors of 28?

5. $\frac{5}{8} = \frac{25}{?}$

6. Draw two similar triangles.

7. $36.7 + 23.4 = ?$

8. Which digit is in the hundredths place in 42.67?

9. Round 6.37 to the nearest tenth.

10. Write $\frac{12}{5}$ as a mixed number.

11. Rename $7\frac{3}{4}$ as an improper fraction.

12. How many inches are in 4 feet?

13. $\frac{7}{9} \bigcirc \frac{9}{10}$

14. Mr. Richards has twice as many clients as Mr. Harris. Mr. Harris has 10 fewer clients than Mrs. Jones. If Mrs. Jones has 35 clients, what is the total number of clients of all three of them?

15. $3\frac{1}{3} + 4\frac{1}{4} = ?$

16. $\frac{15}{20} \times \frac{4}{5} = ?$

17. $\frac{6}{10} \div \frac{2}{10} = ?$

18. Define *median.*

19. $600 \times 70 = ?$

20. Measure the length.

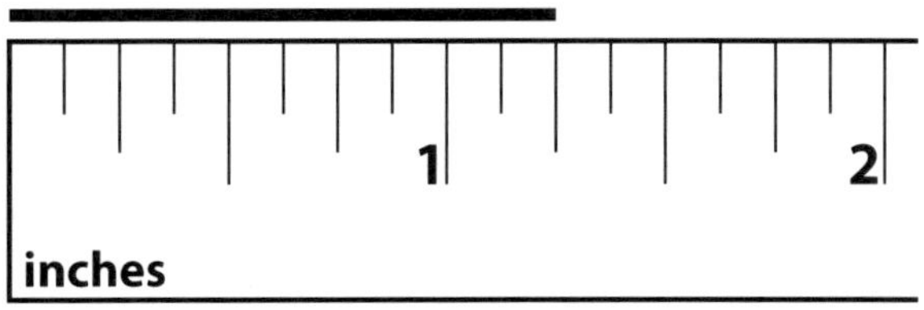

1.	2.	3.	4.
5.	6.	7.	8.
9.	10.	11.	12.
13.	14.	15.	16.
17.	18.	19.	20.

Lesson #74

1. Round 56.32 to the nearest tenth.
2. $6{,}000 - 2{,}347 = ?$
3. Find the range of 55, 59, 60, 59, and 57.
4. Find the mode of the set of numbers in item 3.
5. $53{,}475 + 27{,}986 = ?$
6. $775 \times 8 = ?$
7. Round 36.5 to the nearest whole number.
8. $3.1 \times 0.5 = ?$
9. What is the area of the rectangle?

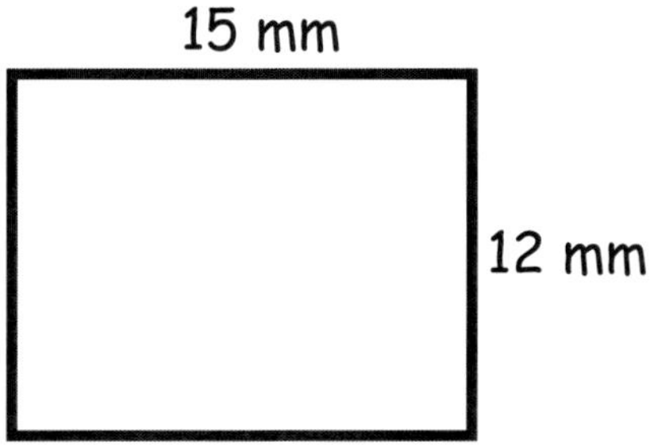

10. $794 \div 5 = ?$
11. If Amy bought 2 CDs for \$14.00 each, a DVD for \$24.99, and a video game for \$37.50, how much change would she get back from a \$100 bill?
12. Find the GCF of 10 and 12.
13. $8\frac{1}{5} \div 6\frac{4}{5} = ?$
14. Which is greater, 200 minutes or 3 hours?
15. Write 16.7 in words.
16. Rename $\frac{15}{4}$ as a mixed number.
17. \$50.00 − \$36.15 = ?
18. $8{,}000 \times 7 = ?$
19. Find the mean of 64, 73, 59, and 76.
20. $\frac{6}{7} \bigcirc \frac{2}{3}$

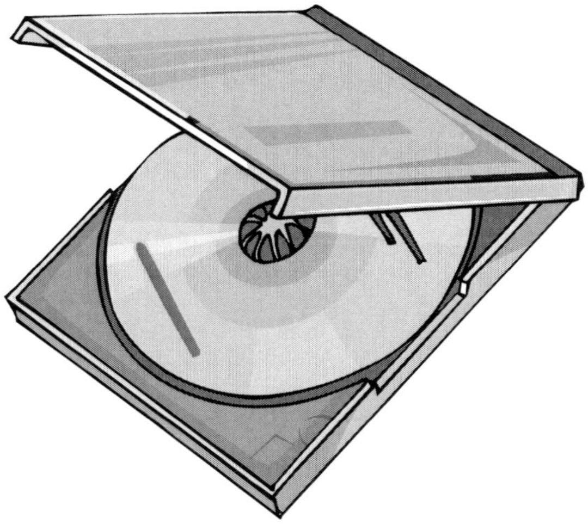

1.	2.	3.	4.
5.	6.	7.	8.
9.	10.	11.	12.
13.	14.	15.	16.
17.	18.	19.	20.

Lesson #75

1. $969 \div 8 = ?$

2. **A *ratio* is a comparison of two numbers. A ratio can be written in 3 forms: word form, fraction form, or ratio form.**

   Write the three ratio forms *2:7*, *2 to 7*, $\frac{2}{7}$ in the box.

3. Find the LCM of 8 and 10.

4. $\frac{3}{6} = \frac{?}{18}$

5. Rename $3\frac{1}{3}$ as an improper fraction.

6. $0.08 \times 0.05 = ?$

7. The answer to a division problem is the __________.

8. $1\frac{1}{3} \times 2\frac{1}{10} = ?$

9. Is the figure to the right a polygon? Explain.

10. Is 42 a prime or a composite number?

11. List the factors of 36.

12. $\frac{12}{15} \times \frac{5}{6} = ?$

13. Put $\frac{12}{24}$ in simplest form.

14. $5\frac{3}{8} - 1\frac{5}{12} = ?$

15. In the fraction $\frac{5}{6}$, which number is in the denominator?

16. Round 7.456 to the nearest hundredth.

17. How many milliliters are in a liter?

18. How many centimeters are in 4 meters?

19. On the Celsius scale, water freezes at _______.

20. $3 \times 6 + 18 = ?$

1.	2.	3.	4.
5.	6.	7.	8.
9.	10.	11.	12.
13.	14.	15.	16.
17.	18.	19.	20.

Lesson #76

1. On the Fahrenheit scale, water freezes at _________.

2. $86 + 24 + 75 = ?$

3. $800 - 655 = ?$

4. Put $\frac{12}{15}$ in simplest form.

5. The answer to an addition problem is called the ________.

6. $\frac{9}{15} \times \frac{5}{6} = ?$

7. Draw a hexagon. How many sides does it have?

8. Which digit is in the tenths place in 36.75?

9. Draw intersecting lines.

10. How many pounds are in 4 tons?

11. Identify the name of the figure shown to the right.

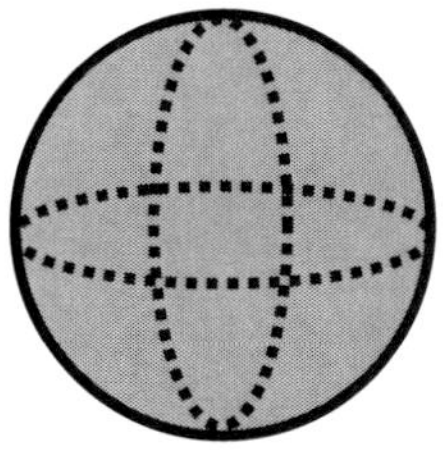

12. $5 \times \frac{2}{3} = ?$

13. Write the ratio *5:7* in two other ways.

14. Find the perimeter of a pentagon if each side measures 6 inches.

15. Find the GCF of 12 and 16.

16. $65 \times 37 = ?$

17. $2\frac{1}{2} + 3\frac{1}{3} = ?$

18. $\frac{2}{3} \times \frac{9}{12} = ?$

Use the stem-and-leaf plot for 19 – 20.

19. What was Gloria's highest score?

20. Find the median of the data.

Gloria's Bean Bag Toss Scores

Stem	Leaf
0	6 9
1	2 8
2	1 4
3	2 5
4	0
5	

1.	2.	3.	4.
5.	6.	7.	8.
9.	10.	11.	12.
13.	14.	15.	16.
17.	18.	19.	20.

Lesson #77

1. Rename $4\frac{2}{3}$ as an improper fraction.
2. Draw intersecting lines.
3. Find the perimeter and the area of the square.

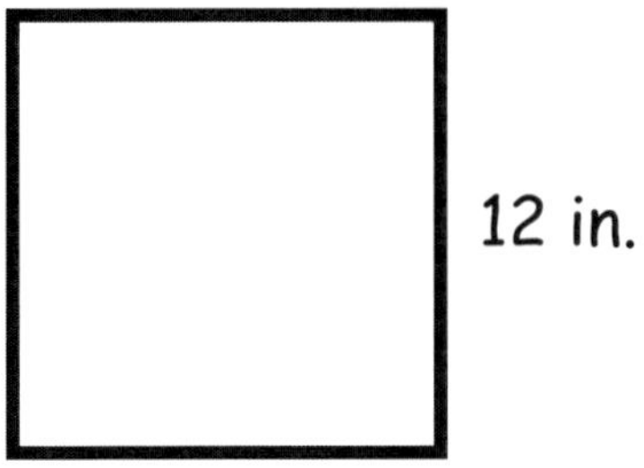

4. $45 \times 23 = ?$
5. What is the value of x? $\frac{4}{5} = \frac{x}{125}$
6. Is the number 21 prime or composite?
7. $652 + 78 + 347 = ?$
8. Mia averages 50 minutes of homework each school night. Usually, Mia has school five days per week; her school year is 39 weeks long. About how many minutes of homework will she do during the school year?
9. $8.3 - 4.75 = ?$
10. Which digit is in the hundredths place in 46.75?
11. How many quarts are in 2 gallons?

12. $\frac{2}{6} \times \frac{12}{14} = ?$
13. $\frac{5}{6} + \frac{2}{3} = ?$
14. Write the ratio *8 to 9* in two other ways.
15. $\frac{7}{8} \bigcirc \frac{5}{6}$
16. Find the LCM of 2, 6, and 8.
17. $15{,}302 - 9{,}475 = ?$
18. It is 4:20. What time will it be in 3 hours and 10 minutes?
19. Round 12.651 to the nearest tenth.
20. Mario is 4 feet 7 inches tall. What is his height in inches?

1.	2.	3.	4.
5.	6.	7.	8.
9.	10.	11.	12.
13.	14.	15.	16.
17.	18.	19.	20.

Lesson #78

1. Draw an acute angle.
2. $5\frac{2}{3}+6\frac{1}{5}=?$
3. 8.25 + 3.4 = ?
4. The figure shown is a parallelogram. 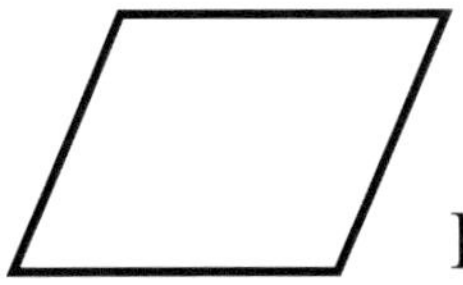 Draw one.
5. $\frac{8}{12}\times\frac{6}{16}=?$
6. Yolanda's school is collecting box tops for cash. Her school will get two cents for every 3 box tops they collected. If the school collects 2,187 box tops, how much money will be generated?
7. 1,000 − 467 = ?
8. (60 + 5) ÷ 5 = ?
9. Write the ratio $\frac{4}{5}$ in two other ways.
10. How many months are there between December 1st and May 1st?
11. List the factors of 36.
12. Round 5.76 to the nearest tenth.
13. 436 × 7 = ?
14. Find the area of this rectangle.

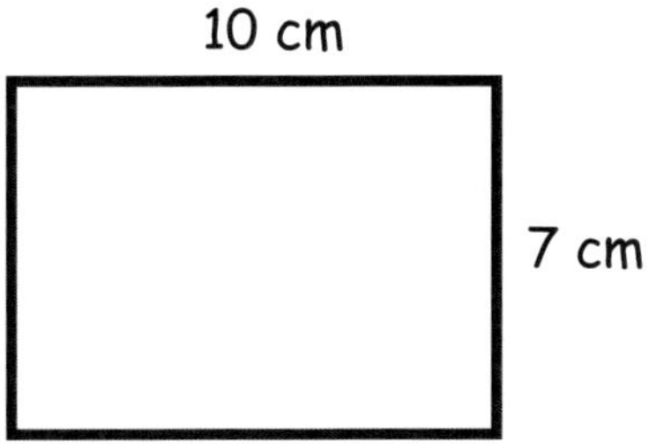

15. Closed shapes made up of line segments are _________.
16. 8,245 ÷ 5 = ?
17. $7\frac{1}{10}+3\frac{3}{10}=?$
18. What year is 2 decades before 1945?
19. How long is the bar in millimeters?

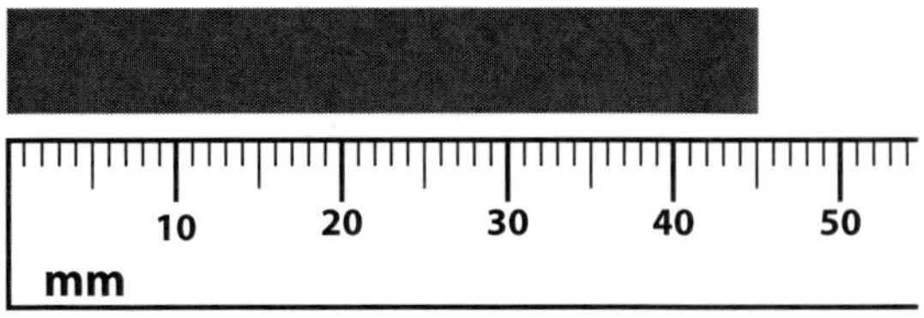

20. $8\frac{1}{7}-3\frac{3}{7}=?$

1.	2.	3.	4.
5.	6.	7.	8.
9.	10.	11.	12.
13.	14.	15.	16.
17.	18.	19.	20.

Lesson #79

1. How many ounces are in 5 pounds?
2. Round 5.3625 to the nearest thousandth.
3. Write the standard number for *thirty-six and fifteen hundredths.*
4. $36 + 25 + 14 + 51 = ?$
5. $60{,}000 - 36{,}478 = ?$
6. At Copies Plus there are 18 copy machines in a row with no space between them. Each copier is 2 feet 3 inches across. Find the total length of the 18 copiers.
7. Find $\frac{1}{6}$ of 42.
8. $92 \times 47 = ?$
9. $3{,}480 \div 40 = ?$
10. $3 - 2\frac{1}{2} = ?$
11. Change $\frac{8}{3}$ to a mixed number.

12. Change $8\frac{3}{4}$ to an improper fraction.
13. Order these decimals from least to greatest. 3.15 3.51 3.105 3.051
14. If a truck can carry $2\frac{1}{2}$ tons, how many pounds can it carry?
15. Write the formula for finding the volume of a prism.
16. $\frac{3}{4} = \frac{?}{16}$
17. Put $\frac{7}{21}$ in simplest form.
18. $3.6 + 2.5 + 1.2 = ?$
19. Define *ratio.*
20. $6\frac{1}{4} + 3\frac{2}{5} = ?$

1.	2.	3.	4.
5.	6.	7.	8.
9.	10.	11.	12.
13.	14.	15.	16.
17.	18.	19.	20.

Lesson #80

1. Draw a line segment.
2. $625 \times 3 = ?$
3. How many feet are in 2 miles?
4. $8{,}020 - 2{,}765 = ?$
5. List the factors of 18.
6. How many cups are in a pint?
7. $\frac{3}{4} \times \frac{6}{18} = ?$
8. How many milliliters are in a liter?
9. Solve the proportion for x. $\frac{5}{7} = \frac{x}{91}$
10. $6\frac{1}{3} + 1\frac{2}{5} = ?$
11. $4\frac{2}{5} - 2\frac{3}{5} = ?$
12. Find $\frac{2}{3}$ of 120.
13. $8{,}652 \div 45 = ?$
14. $\frac{7}{8} \bigcirc \frac{5}{6}$
15. Find the GCF of 12 and 36.
16. Write the ratio *6:7* in two other ways.
17. Put $\frac{9}{12}$ in simplest form.
18. Draw a square. Show two lines of symmetry.
19. What is the mode of the math scores?
20. What is the range of the scores?

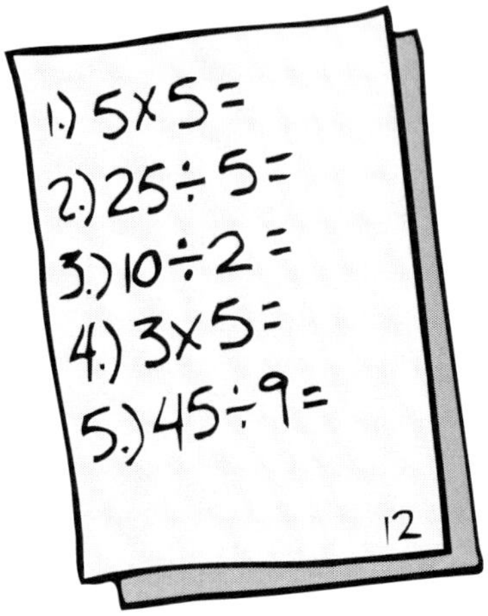

Math Scores for Room 202

Stem	Leaf
5	4
6	8
7	2 6 9 9
8	4 6 8 8 8
9	0 3 5 6 6 8 9

1.	2.	3.	4.
5.	6.	7.	8.
9.	10.	11.	12.
13.	14.	15.	16.
17.	18.	19.	20.

Lesson #81

1. Draw an obtuse angle.
2. The diameter of a circle is 16 mm. What is its radius?
3. $6.2 \times 0.7 = ?$
4. $3.5 \div 5 = ?$
5. Is 29 a prime or a composite number?
6. $862 + 479 = ?$
7. $6\frac{1}{3} - 3\frac{2}{3} = ?$
8. Find the mean and the median of 62, 60, 57, 56, and 65.
9. $70{,}000 - 34{,}973 = ?$
10. $28.8 \div 0.6 = ?$
11. For each item, would a milliliter or a liter be the more reasonable unit of measure?
    A) aquarium B) soup can C) eye dropper
12. $\frac{9}{11} \bigcirc \frac{7}{12}$
13. Round 5.276 to the nearest hundredth.
14. At a kennel, the ratio of dogs to cats was 4 to 9. If there were 64 dogs, how many cats were at the kennel?
15. Find the LCM of 9 and 21.
16. Put these numbers in order from least to greatest.
    3.652 3.65 3.562 3.056

17. Rename $4\frac{1}{2}$ as an improper fraction.
18. $2\frac{2}{3} \times 3\frac{1}{4} = ?$
19. At what temperature does water boil in degrees Fahrenheit?
20. How many inches are in 7 feet?

1.	2.	3.	4.
5.	6.	7.	8.
9.	10.	11.	12.
13.	14.	15.	16.
17.	18.	19.	20.

Lesson #82

1. $6\frac{2}{3} \times 3\frac{1}{5} = ?$
2. How many quarts are in 6 gallons?
3. Find the LCM of 12 and 20.
4. Solve the proportion for x. $\frac{5}{9} = \frac{x}{117}$
5. 8,463 + 24,592 = ?
6. 36 × 14 = ?
7. $\frac{4}{5} \bigcirc \frac{11}{12}$
8. 63 + ? = 97
9. How many days are in 12 weeks?
10. The radius of a circle is 14 cm. What is the circle's diameter?
11. Find the perimeter and the area of this rectangle.

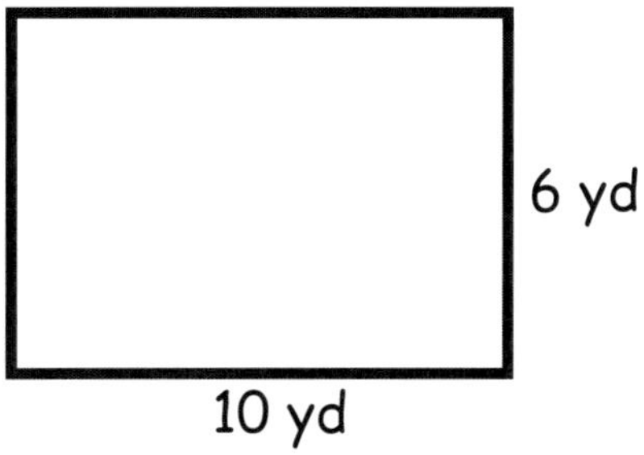

12. Find the median of 48, 72, 83, 49, and 16.
13. Write $\frac{91}{9}$ as a mixed number.
14. Write the ratio *4 to 9* in two other ways.
15. Draw a pentagon. Draw a line of symmetry.
16. Figures with the same shape but different sizes are ________.
17. 580 ÷ 32 = ?
18. $10\frac{1}{5} - 3\frac{3}{10} = ?$
19. Find the mean of 80, 82, 88, 30, and 45.
20. Goodwin School's fifth graders left their school at 9:30 a.m. The bus ride to the museum was 1 hour and 15 minutes. The museum tour lasted 3 hours; afterwards the students spent 50 minutes eating lunch. At what time did they board the bus to return to school?

1.	2.	3.	4.
5.	6.	7.	8.
9.	10.	11.	12.
13.	14.	15.	16.
17.	18.	19.	20.

Lesson #83

1. Solve the proportion to find x. $\frac{4}{7} = \frac{56}{x}$

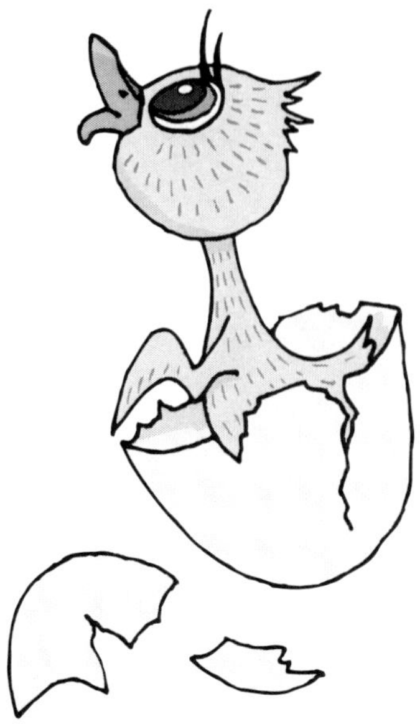

2. $62.5 + 37.82 = ?$
3. Rename $\frac{37}{6}$ as a mixed number.
4. Round 6.575 to the nearest hundredth.
5. $5\frac{2}{5} \times \frac{5}{9} = ?$
6. How many cups are in 4 pints?
7. $6.2 \times 4.1 = ?$
8. Is 17 a prime or a composite number?
9. Find the LCM of 15 and 25.
10. On the Celsius temperature scale, water boils at __________.
11. How many centimeters are in 5 meters?
12. Find the median of 12, 89, 16, 28, and 76.
13. $0.09 \times 0.5 = ?$
14. Find the volume of the prism.

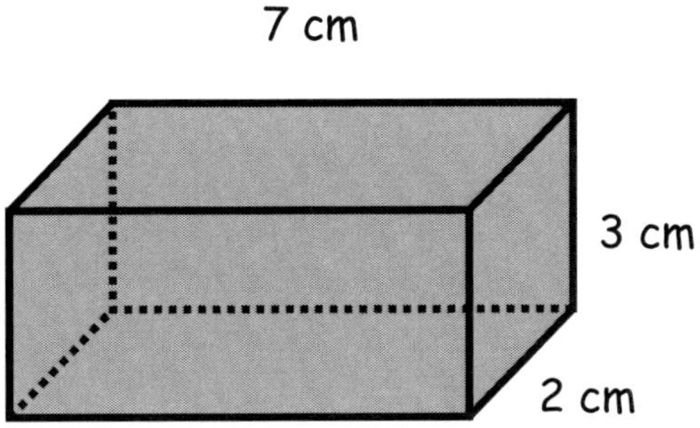

15. $(50 \times 6) \times 2 = ?$
16. Draw a right angle.
17. Write 40% as a fraction in simplest form.
18. $9{,}860 \div 4 = ?$
19. $4\frac{5}{8} - 2\frac{1}{4} = ?$
20. What is the probability that the spinner will stop on each?
    A) the number 2
    B) an odd number
    C) a number less than 5

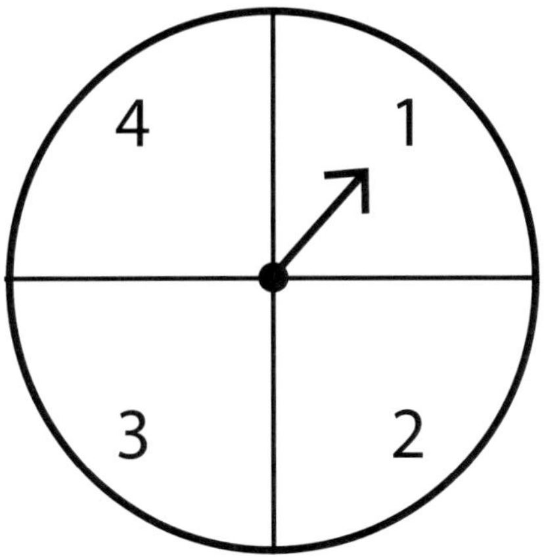

1.	2.	3.	4.
5.	6.	7.	8.
9.	10.	11.	12.
13.	14.	15.	16.
17.	18.	19.	20.

Lesson #84

1. Define *ratio.*

2. Jeff is 6 feet 2 inches tall. Give Jeff's height in inches.

3. $405 \times 16 = ?$

4. $3.4 - 1.72 = ?$

5. Round 51,623,472 to the nearest ten million.

6. Put $\frac{15}{20}$ in simplest form.

7. Find the LCM of 6, 9, and 10.

8. $\frac{7}{9} \bigcirc \frac{2}{5}$

9. $402 - 176 = ?$

10. Write 25% as a reduced fraction and a decimal.

11. If it is 12:25 now, what time was it 3 hours and 10 minutes ago?

12. $345 + 68 + 9 = ?$

13. $\frac{2}{3} \times \frac{5}{6} = ?$

14. Write 7.2 using words.

15. Between which 2 years did the greatest increase in sales occur? In which year did sales decrease?

16. How much money did Shoe World make over the five year period?

17. How many pounds are in 6 tons?

18. Water freezes at ________ degrees Celsius.

19. The diameter of a circle is 18 centimeters. What is the radius?

20. Find the median of 86, 24, 13, 73, and 57.

1.	2.	3.	4.
5.	6.	7.	8.
9.	10.	11.	12.
13.	14.	15.	16.
17.	18.	19.	20.

Lesson #85

1. What is the value of x? $\frac{3}{8} = \frac{x}{80}$
2. $863 + 479 = ?$
3. Find the mean of 27, 36, and 15.
4. A truck is 48 ft long and 10 ft wide. The owner wants to paint a red stripe around the outside of the truck. How long is the stripe?

5. $3.2 \times 4.7 = ?$
6. $5.25 \div 3 = ?$
7. Find the GCF of 6 and 10.
8. $\frac{1}{4} + \frac{2}{5} = ?$
9. $5\frac{1}{3} - 2\frac{2}{3} = ?$
10. $2.772 \div 2.1 = ?$
11. Find the volume of the rectangular prism.

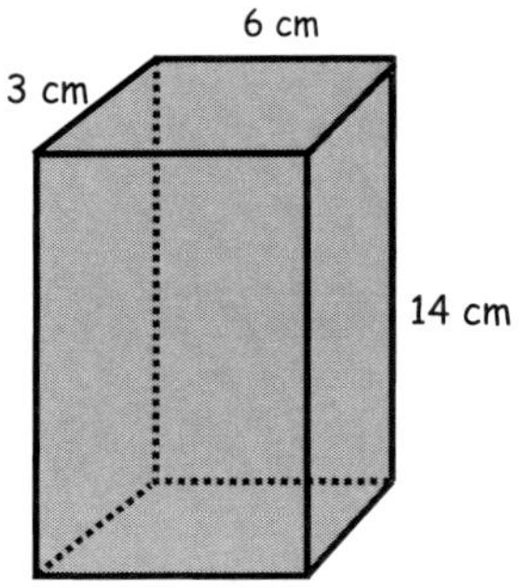

12. $\frac{3}{6} \times \frac{2}{9} = ?$
13. $\frac{3}{7} = \frac{?}{21}$
14. Which is the greater volume, 60 milliliters or 40 liters?
15. Write 76.34 using words.
16. Sharon worked $3\frac{3}{4}$ hours on Tuesday and $6\frac{1}{2}$ hours on Wednesday. How many hours did she work in all?
17. Write the ratio $\frac{2}{9}$ in two other ways.
18. How many feet are in a mile?
19. Round 436,756,213 to the nearest hundred million.
20. What is the probability that the spinner will land in the area marked A?

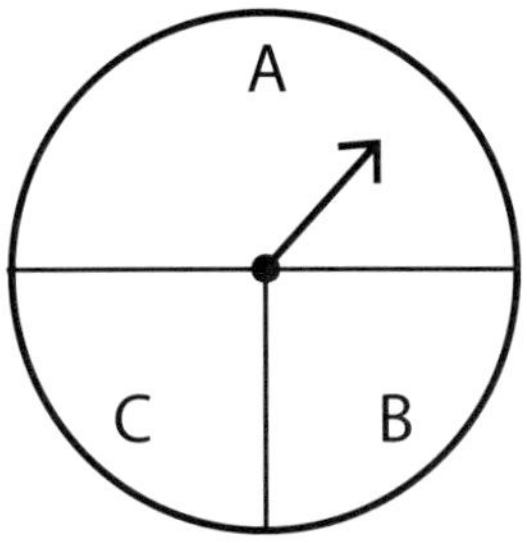

1.	2.	3.	4.
5.	6.	7.	8.
9.	10.	11.	12.
13.	14.	15.	16.
17.	18.	19.	20.

Lesson #86

1. $56.2 \times 0.06 = ?$
2. How many years are 5 decades?
3. Write 0.48 as a reduced fraction and a percent.
4. $7.6 - 2.75 = ?$
5. Draw an octagon. Label a side 7 inches long. Find the perimeter.
6. Find the GCF and the LCM of 12 and 14.
7. Which digit is in the thousandths place in 32.076?
8. Figures with the same shape and the same size are _______.
9. Put $\frac{12}{14}$ in simplest form.
10. $1,862 + 9,754 = ?$
11. $5\frac{1}{2} + 3\frac{1}{4} = ?$
12. How many ounces are in a pound?
13. Round 4.675 to the nearest tenth.
14. $60,000 - 39,417 = ?$
15. $4.35 \div 0.5 = ?$

16. The ratio of flowers to weeds in the garden is 5 to 7. If there are 30 flowers, how many weeds are there?
17. Write $\frac{47}{5}$ as a mixed number.
18. $\frac{3}{4} \div \frac{1}{2} = ?$
19. $4\frac{2}{5} - 2\frac{4}{5} = ?$
20. Use the pie graph to answer the following:

    A) What percent of students got an A in math?
    B) What percent got a B in math?
    C) What percent got a C or a D in math?

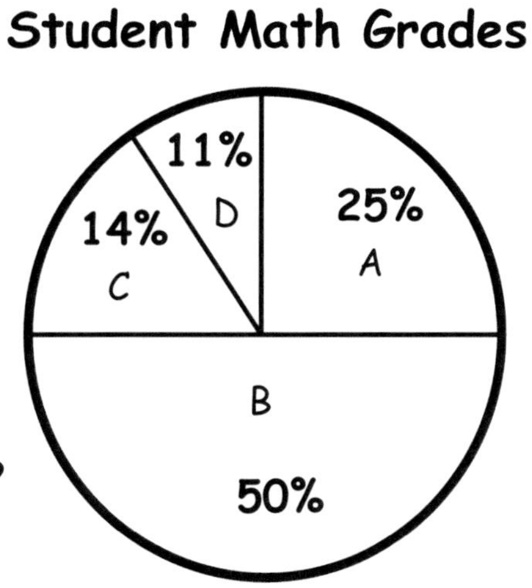

1.	2.	3.	4.
5.	6.	7.	8.
9.	10.	11.	12.
13.	14.	15.	16.
17.	18.	19.	20.

Lesson #87

1. $\frac{3}{8} \times \frac{12}{15} = ?$
2. How many centimeters are in 9 meters?
3. $6.75 \div 0.5 = ?$
4. $7\frac{1}{5} - 3\frac{3}{5} = ?$
5. Find the LCM of 10 and 15.
6. $\frac{6}{7} \bigcirc \frac{5}{6}$
7. Is 81 a prime or a composite number?
8. Write 30% as a reduced fraction and a decimal.
9. Find the area of a rectangle that is 12 feet long and 6 feet wide.
10. $8\frac{1}{2} + 6\frac{3}{4} = ?$
11. The ratio of girls to boys at the park is $\frac{5}{6}$. If there are 35 girls, how many boys are at the park?
12. The diameter of a circle is 14 in. What is the radius?
13. Change $3\frac{1}{3}$ to an improper fraction.
14. How many pounds are in $4\frac{1}{2}$ tons?
15. $63 \times 47 = ?$
16. List the factors of 24.
17. Draw a parallelogram.
18. What is the range of afternoon temperatures?
19. What is the mode of these temperatures?
20. Round 8,762 to the nearest ten.

Summer Afternoon Temperatures

Stem	Leaf
6	3 3 4
7	0 0 0 7 9
8	4 5 5 7 8
9	2 3 3

1.	2.	3.	4.
5.	6.	7.	8.
9.	10.	11.	12.
13.	14.	15.	16.
17.	18.	19.	20.

Lesson #88

1. A major snowstorm canceled two-thirds of the flights out of the airport. If there were 471 flights scheduled that day, how many flights were cancelled? How many flights left the airport?

2. Round 56.47 to the nearest whole number.

3. $\frac{3}{5}+\frac{2}{5}=?$

4. Draw intersecting lines.

5. Write 67.25 in words.

6. $\frac{5}{6}\times\frac{12}{15}=?$

7. $365+476+213=?$

8. $800-154=?$

9. $3\frac{1}{6}-2\frac{5}{6}=?$

10. Write the ratio *8 to 9* in two other ways.

11. How many inches are in a yard?

12. A quadrilateral has _____ sides.

13. Define *percent.*

14. $325\times 14=?$

15. $6.2+4.75=?$

16. List the factors of 18.

17. $5\frac{2}{3}+3\frac{1}{6}=?$

18. $4.73-2.965=?$

19. Put these decimals in order from least to greatest.

    4.62 4.6 4.0 4.623

20. Write $\frac{15}{4}$ as a mixed number.

1.	2.	3.	4.
5.	6.	7.	8.
9.	10.	11.	12.
13.	14.	15.	16.
17.	18.	19.	20.

Lesson #89

1. $5.2 - 3.64 = ?$
2. Write 0.45 as a reduced fraction and a percent.
3. Which digit is in the tenths place in 39.47?
4. Put these decimals in order from greatest to least.

   2.07 2.007 2.0 2.70
5. Put $\frac{14}{16}$ in simplest form.
6. $573 \times 9 = ?$
7. Write $6\frac{2}{5}$ as an improper fraction.
8. $782 + 463 + 219 = ?$
9. How many days are in 6 weeks?
10. Find the perimeter and the area of the square.

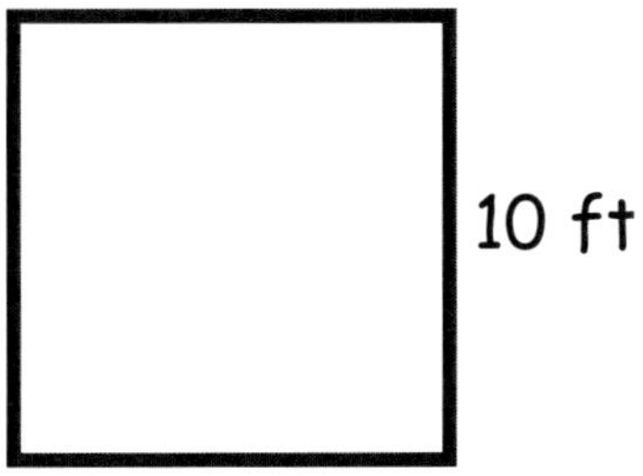

11. $\frac{5}{6} = \frac{?}{36}$
12. $7,000 - 2,643 = ?$
13. Find the GCF of 18 and 24.
14. It is 12:35. What time was it 3 hours and 35 minutes ago?
15. Find $\frac{2}{5}$ of 20.
16. List the factors of 12.
17. Darnell ran 550 yards in 3 minutes. At this rate, how many yards can he run in 6 minutes?
18. $1,200 \div 38 = ?$
19. What part is shaded? Write your answer as a decimal and as a fraction.

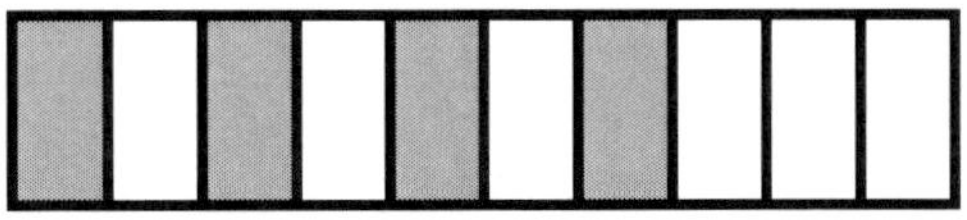

20. $3 \times \frac{5}{6} = ?$

1.	2.	3.	4.
5.	6.	7.	8.
9.	10.	11.	12.
13.	14.	15.	16.
17.	18.	19.	20.

Lesson #90

1. $182 \times 15 = ?$
2. Solve the proportion for *x*. $\frac{4}{5} = \frac{x}{75}$
3. Define *diameter.*
4. $4\frac{2}{7} - 2\frac{5}{7} = ?$
5. $\frac{18}{16} \times \frac{12}{16} = ?$
6. $5.1 - 3.72 = ?$
7. $\frac{5}{6} \bigcirc \frac{2}{3}$
8. $\frac{4}{10} \div \frac{2}{10} = ?$
9. Write 75% as a decimal.
10. Find the volume of the rectangular prism.

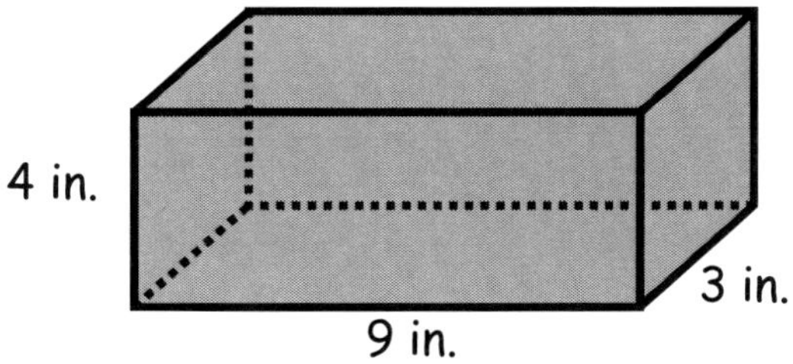

11. Write the number 56.897 in words.
12. $76{,}908 + 34{,}775 = ?$
13. $(400 + 500) + 300 = ?$
14. Find the value of *x*. $\frac{3}{10} = \frac{x}{100}$
15. $5{,}090 - 2{,}765 = ?$
16. $16.8 \times 1.2 = ?$
17. $9.52 \div 0.28 = ?$
18. $1\frac{2}{3} \div \frac{7}{8} = ?$
19. There are 30 students in the class. How many have blonde hair?
20. How many have black hair? Brown hair?

Student Hair Color

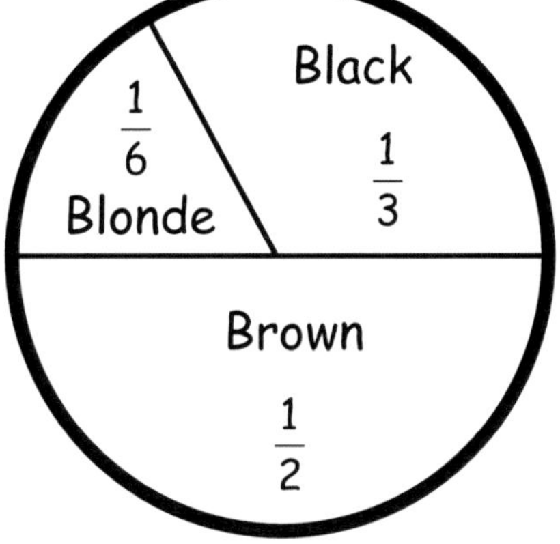

1.	2.	3.	4.
5.	6.	7.	8.
9.	10.	11.	12.
13.	14.	15.	16.
17.	18.	19.	20.

Lesson #91

1. Round \$45.67 to the nearest dollar.
2. What time will it be in 8 hours and 15 minutes if it is 12:30 now?
3. $52.92 \div 2.1 = ?$
4. $45 + 89 + 135 = ?$
5. $3 \times 5 \times 7 \times 2 = ?$
6. Find the area of the rectangle.

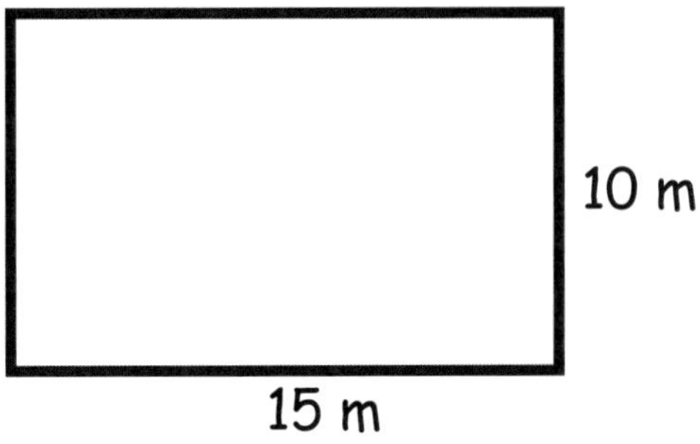

7. $4.56 - 1.3 = ?$
8. Write the standard number for *sixteen and thirty-six hundredths.*
9. Find the value of x. $\frac{3}{8} = \frac{x}{120}$
10. $20{,}000 - 12{,}703 = ?$
11. Round 7.035 to the nearest hundredth.
12. $\frac{3}{2} \div \frac{2}{3} = ?$
13. $5\frac{1}{7} + 7\frac{2}{3} = ?$
14. Write $\frac{17}{3}$ as a mixed number.
15. Find the LCM of 15 and 25.

16. An elevator can hold a maximum of 6 people or 1200 pounds. Grace weighs 60 pounds; her dad weighs three times her weight, and her brother weighs twice her weight. Grace's mom weighs 30 pounds more than Grace and her sister is 50 pounds heavier. Can they all ride in the elevator together? Explain your answer.
17. $\frac{12}{18} \times \frac{9}{10} = ?$
18. $3\frac{1}{5} - 1\frac{4}{5} = ?$
19. List the factors of 15.
20. What is the length of the bar?

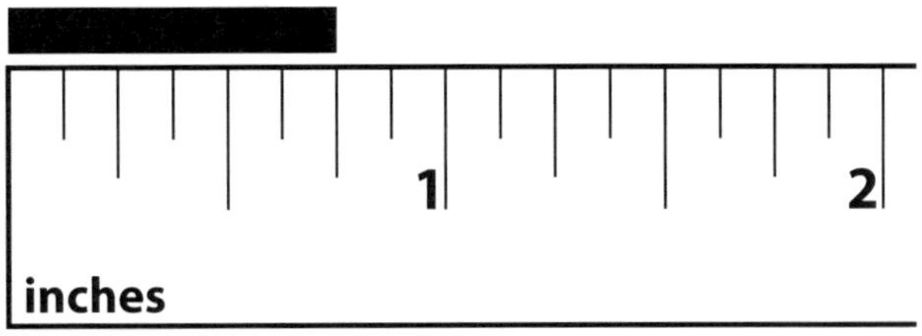

1.	2.	3.	4.
5.	6.	7.	8.
9.	10.	11.	12.
13.	14.	15.	16.
17.	18.	19.	20.

Lesson #92

1. What word means *per hundred*?
2. Find $\frac{1}{7}$ of 56.
3. $\frac{2}{3} \div \frac{1}{4} = ?$
4. 359 + 981 = ?
5. How many ounces are in 5 pounds?
6. $0.735 \times 0.5 = ?$
7. $4.3 - 1.21 = ?$
8. Find the GCF of 8 and 24.
9. $\frac{2}{3} - \frac{1}{6} = ?$
10. Grandma has 25 blankets. She gave two-fifths of them away. How many blankets does she have left?
11. Find the perimeter of a square whose sides measure 12 yards each.
12. Find the mean of 93, 89, and 97.
13. How many centimeters are in 7 meters?
14. Which digit is in the ten thousands place in the number 624,798?
15. Write $\frac{4}{5}$ as a percent.
16. Solve for x. $\frac{2}{5} = \frac{x}{15}$
17. Put $\frac{10}{12}$ in simplest form.
18. \$100 − \$88.74 = ?
19. List the factors of 24.
20. $5 - 1\frac{2}{5} = ?$

1.	2.	3.	4.
5.	6.	7.	8.
9.	10.	11.	12.
13.	14.	15.	16.
17.	18.	19.	20.

Lesson #93

1. Round 34.89 to the nearest tenth.
2. Write the standard number for *six and fourteen hundredths.*
3. 40,000 − 24,786 = ?
4. $4\frac{1}{4} - 2\frac{3}{4} = ?$
5. 3.67 × 0.5 = ?
6. Draw a pentagon. Show a line of symmetry.
7. 208.4 ÷ 8 = ?

8. The Thomas family spent $4.95 each for 2 adult movie tickets and $3.50 each for 3 children's tickets. They spent $17.98 on snacks and $4.50 on parking. How much did the Thomas family spend?
9. $1\frac{4}{9} + 3\frac{2}{3} = ?$
10. Put the decimals in order from greatest to least.
    3.7 3.07 3.77 3.077
11. Which factors of 20 are also factors of 45?
12. Draw perpendicular lines.
13. Eight weeks are how many days?
14. 897 + 653 = ?
15. $4 \times \frac{2}{3} = ?$
16. It is 3:10. What time will it be in 6 hours and 5 minutes?
17. Solve for *x*. $\frac{4}{5} = \frac{x}{65}$
18. Write 65% as a reduced fraction.
19. Write the ratio *4:7* in two other ways.
20. What part is shaded? Write your answer as a decimal and as a fraction.

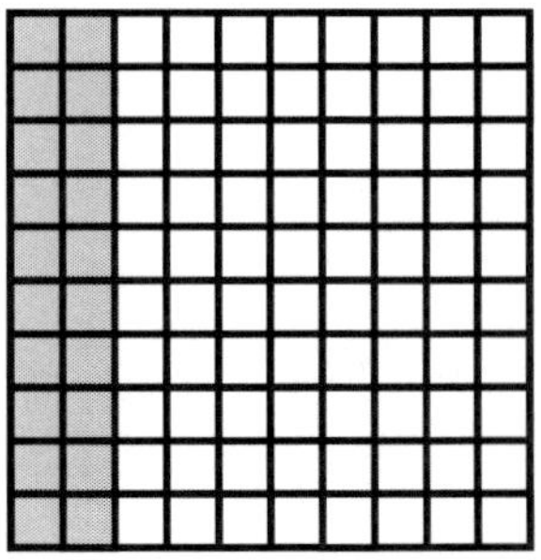

1.	2.	3.	4.
5.	6.	7.	8.
9.	10.	11.	12.
13.	14.	15.	16.
17.	18.	19.	20.

Lesson #94

1. How many minutes are in 6 hours?

2. $2.5 \times 3.6 = ?$

3. The heart of a basset hound pumps 10.4 liters of blood in 5 minutes. How many liters of blood per minute is this?

4. Make a factor tree for 32.

5. $6 - 2\frac{7}{8} = ?$

6. $468 + 988 = ?$

7. Write 56.79 in words.

8. Which is greater, 45 feet or 12 yards?

9. Round 4,875,908 to the nearest ten thousand.

10. $55 \times 27 = ?$

11. How many centimeters are in 9 meters?

12. $\frac{4}{5} \bigcirc \frac{2}{3}$

13. $\frac{4}{5} + \frac{7}{8} = ?$

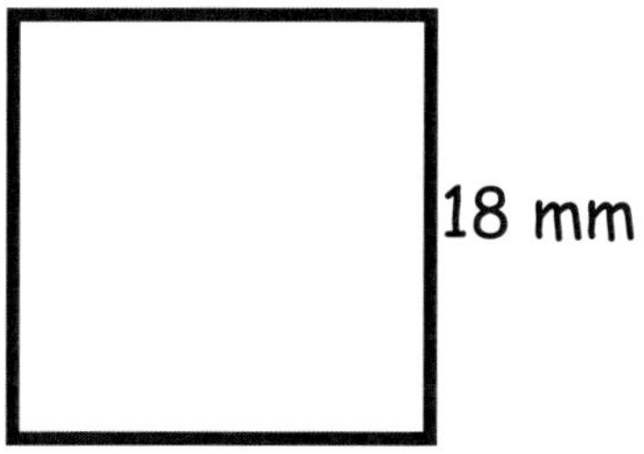

14. Find the perimeter of the square.

15. On the Celsius temperature scale, water freezes at __________.

16. $56.78 + 9.34 = ?$

17. How many pounds are 80 ounces?

18. Write 45% as a decimal and as a reduced fraction.

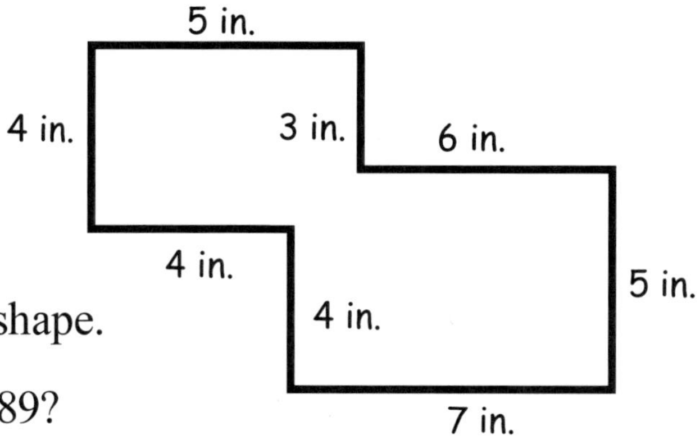

19. Find the perimeter of this irregular shape.

20. What is the mean of 229, 393, and 389?

1.	2.	3.	4.
5.	6.	7.	8.
9.	10.	11.	12.
13.	14.	15.	16.
17.	18.	19.	20.

Lesson #95

1. What is the sixth prime number?
2. 3,975,442 + 5,799,523 = ?
3. Find the product of 0.56 and 2.4.
4. How many pounds are $3\frac{1}{2}$ tons?
5. Find $\frac{3}{5}$ of 20.
6. 18,000 − 6,988 = ?
7. Figures that have the same shape and the same size are __________.
8. Which digit is in the thousandths place in 334.887?
9. The 37 students in Mrs. Greer's class bought their teacher a birthday gift that cost $464.72. How much did each student have to contribute to divide the cost evenly?
10. Find the GCF and LCM of 16 and 24.
11. $10\frac{1}{6} - 3\frac{5}{6} = ?$
12. Which is greater, 0.78 or 0.708?

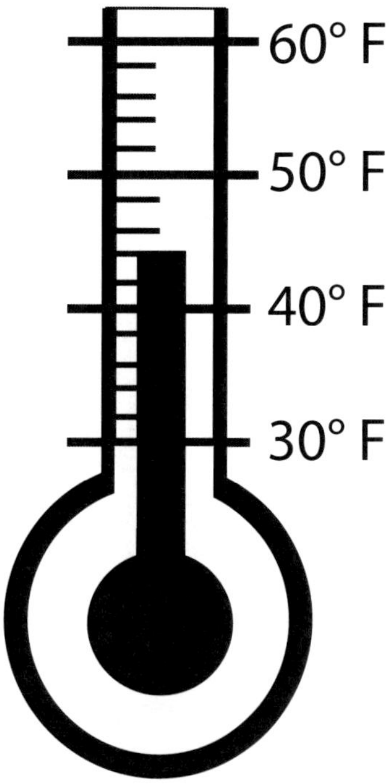

13. $\frac{5}{6} \times \frac{12}{15} = ?$
14. Write 0.36 as a percent and as a reduced fraction.
15. 35.6 + 98.67 = ?
16. 50 × 5 ÷ 25 = ?
17. $\frac{5}{9}$ ◯ $\frac{7}{10}$
18. Find the mode of the numbers 50, 37, 88, 50, and 67.
19. What temperature is shown on the thermometer?
20. Water freezes at 32°F. The temperature on this thermometer is how many degrees above freezing?

1.	2.	3.	4.
5.	6.	7.	8.
9.	10.	11.	12.
13.	14.	15.	16.
17.	18.	19.	20.

Lesson #96

1. Fran's MP3 player can hold 20 hours of music. She has thirty 15-minute songs, fifteen 3-minute songs, and twenty-five 20 minute songs on it already. How much time does she have left to fill?

2. How many pounds are 400 ounces?

3. $45.8 \times 2.6 = ?$

4. $54.08 \div 0.4 = ?$

5. $6 - 2\frac{2}{3} = ?$

6. How many seconds are in 4 hours?

7. Make a factor tree for 36.

8. $\frac{4}{5} + \frac{5}{8} = ?$

9. Find the area of a square if a side measures 15 millimeters.

10. Solve to find the value of x. $\frac{8}{9} = \frac{120}{x}$

11. What is the answer to a multiplication problem called?

12. Round 14.938 to the nearest tenth.

13. What time was it 3 hours and 15 minutes ago, if it is 5:20 now?

14. How many milliliters are in 4 liters?

15. $\frac{5}{11} \div \frac{15}{22} = ?$

16. Which is less, 0.561 or 0.516?

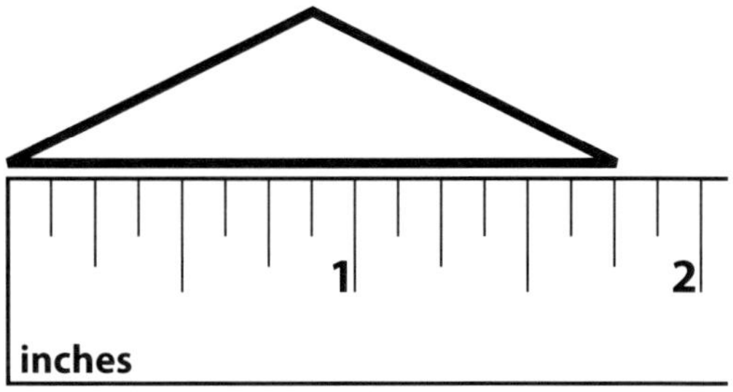

17. Find the median of 13, 78, 56, 44, and 21.

18. Write $6\frac{5}{10}$ as a decimal.

19. How long, in inches, is the base of the triangle?

20. Three out of twenty-five students received an A on the Science test. What fraction of the students received an A on the test?

1.	2.	3.	4.
5.	6.	7.	8.
9.	10.	11.	12.
13.	14.	15.	16.
17.	18.	19.	20.

Lesson #97

1. One hour and ten minutes is how many seconds?

2. How many feet are in 4 miles?

3. Write the name of this shape.

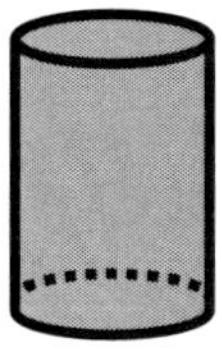

4. $45.79 - 23.6 = ?$

5. List the factors of 20.

6. $4\frac{5}{6} + 3\frac{2}{3} = ?$

7. How many inches are in 6 yards?

8. On the Celsius scale, water freezes at what temperature?

9. Find $\frac{2}{7}$ of 35.

10. What is the value of x? $\frac{5}{9} = \frac{x}{135}$

11. $3{,}070 - 1{,}224 = ?$

12. The distance across a circle is called the diameter. What do we call the distance around the outside of a circle?

13. Write $\frac{4}{5}$ as a decimal and as a percent.

14. The radius of a circle is 50 cm. What is the diameter?

15. Rename $\frac{8}{6}$ as a mixed number.

16. $997 \div 53 = ?$

17. The ratio of ducks to geese on the pond is 3:7. If there are 39 ducks, how many geese are on the pond?

18. $15 - 8\frac{1}{6} = ?$

19. Is the angle obtuse or acute?

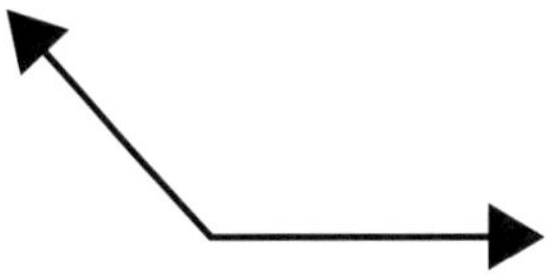

20. What is the probability that the spinner will stop on an even number? On an odd number? On a number less than 4?

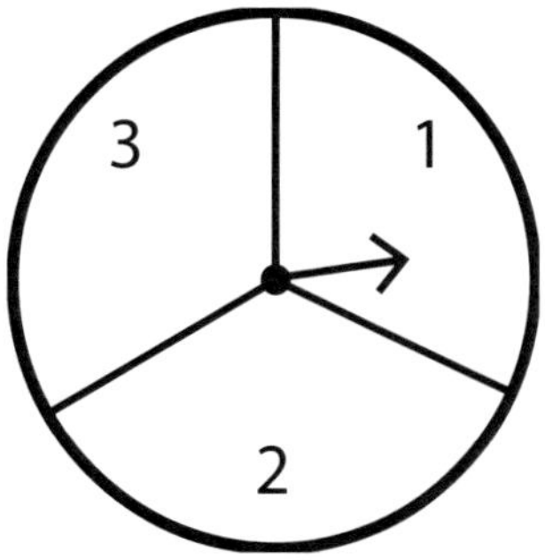

1.	2.	3.	4.
5.	6.	7.	8.
9.	10.	11.	12.
13.	14.	15.	16.
17.	18.	19.	20.

Lesson #98

1. $342 \times 25 = ?$

2. On the first day of practice Tom did 3 sit-ups. The next day, he did 5. On day 3 he did 8, and on day 4 he did 12 sit-ups. If this pattern continues, how many sit-ups will Tom do on day 6?

3. Write the formula for finding the volume of a prism.

4. $28{,}897 \div 9 = ?$

5. Write the standard number for *twelve and three hundredths.*

6. $7 - 3\frac{1}{3} = ?$

7. Write 0.85 as a percent and as a reduced fraction.

8. $\frac{3}{8} \bigcirc \frac{5}{6}$

9. Write 104.87 in words.

10. The ratio of oranges to apples is 5 to 6. If there are 84 apples, how many oranges are there?

11. Round 4,776,223 to the nearest hundred thousand.

12. $\frac{5}{6} \times \frac{10}{15} = ?$

13. A ball weighs 2 pounds, 3 ounces. What is the weight in ounces?

14. How many inches are in 2 yards?

15. $\frac{12}{2} \div \frac{1}{2} = ?$

16. Put $\frac{8}{32}$ in simplest form.

17. $15\frac{2}{3} + 12\frac{1}{6} = ?$

18. Which is greater, 0.009 or 0.097?

19. $456 + 26 + 998 = ?$

20. Solve the proportion for x. $\frac{2}{7} = \frac{x}{91}$

1.	2.	3.	4.
5.	6.	7.	8.
9.	10.	11.	12.
13.	14.	15.	16.
17.	18.	19.	20.

Lesson #99

1. $4\frac{1}{3} - 1\frac{2}{3} = ?$

2. Bill bought 7 pounds of cherries for \$3.43. What was the price for a pound of cherries?

3. Mary's height is 5 feet 2 inches. How many inches tall is Mary?

4. $37.6 + 9.8 + 23.45 = ?$

5. What fraction is shaded?

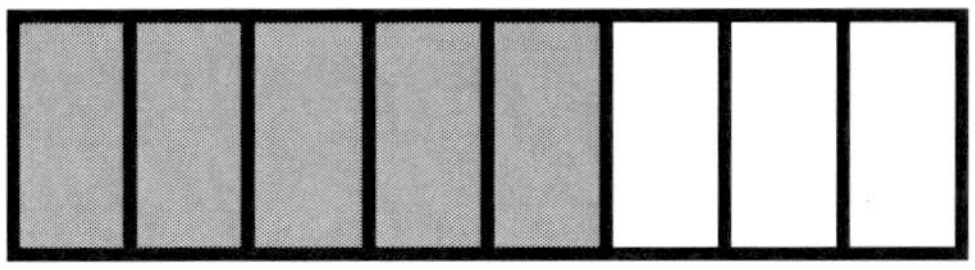

6. How many yards are in one half of a mile?

7. Which digit is in the tenths place in 245.786?

8. Order these decimals from least to greatest.
   3.08 3.8 3.008 3.0

9. What is $\frac{3}{4}$ of 16?

10. $14.58 \div 0.6 = ?$

11. How many faces does a cube have?

12. Estimate the product of 632 and 470 by rounding to the nearest hundred.

13. $\frac{4}{6} \times \frac{15}{16} = ?$

14. Write $\frac{7}{25}$ as a decimal and as a percent.

15. The radius of a circle is 8 inches. What is the diameter?

16. Put $\frac{8}{10}$ in simplest form.

17. What is the shape of a can of peas?

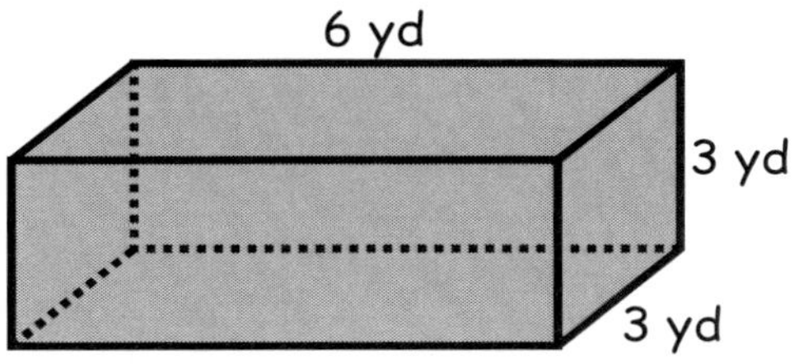

18. $\frac{15}{6} \times \frac{5}{6} = ?$

19. Find the volume of this rectangular prism.

20. The perimeter of a square is 24 inches. How long is each side?

1.	2.	3.	4.
5.	6.	7.	8.
9.	10.	11.	12.
13.	14.	15.	16.
17.	18.	19.	20.

Lesson #100

1. $5{,}091 - 1{,}677 = ?$

2. $400 + 300 - 250 = ?$

3. $3.1 \times 2.3 = ?$

4. $\frac{3}{5} \times \frac{5}{9} = ?$

5. Write 65% as a decimal and as a reduced fraction.

6. Shawna can run 660 yards in 3 minutes. At this rate, how many yards could she run in 9 minutes?

7. $9\frac{6}{7} - 2\frac{1}{14} = ?$

8. Solve for x. $\frac{3}{5} = \frac{x}{75}$

9. Find the GCF and LCM of 10 and 15.

10. Make a factor tree for 64.

11. Round $7.54 to the nearest dollar.

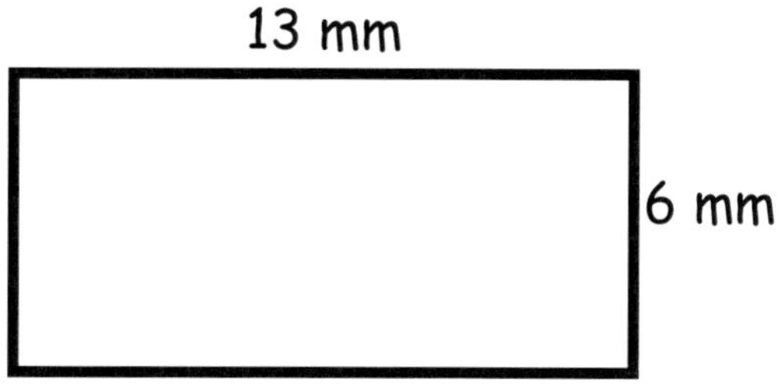

12. $2.6 + 9.87 + 5.1 = ?$

13. How many cups are in 4 pints?

14. Find the area of the rectangle.

15. How many teaspoons are in 3 tablespoons?

16. Nate worked on his homework from 3:30 p.m. to 5:00 p.m. For how many minutes did he work on his homework?

17. $3\frac{5}{8} + 1\frac{7}{8} = ?$

18. Write the formula for finding the perimeter of any polygon.

19. The answer to a subtraction problem is the __________.

20. $0.300 \div 4 = ?$

1.	2.	3.	4.
5.	6.	7.	8.
9.	10.	11.	12.
13.	14.	15.	16.
17.	18.	19.	20.

Lesson #101

1. $356 + 982 = ?$
2. Find $\frac{1}{6}$ of 42.
3. How many days are in 2 years?
4. $3\frac{4}{5} + 2\frac{1}{4} = ?$
5. Round 823,788,665 to the nearest ten million.
6. Fairview Elementary students went on a bowling trip. Of the 400 students, $\frac{1}{5}$ had their own bowling shoes. How many students needed to rent shoes?
7. $45.6 - 9.87 = ?$
8. $\frac{4}{7} \times \frac{14}{16} = ?$
9. Which is greater, 0.65 or 0.065?
10. Is 19 a prime or a composite number?
11. Write the ratio *3:8* in two other ways.
12. How many feet are in 3 miles?
13. Write $3\frac{7}{10}$ as a decimal.
14. $\frac{9}{10} \bigcirc \frac{8}{9}$
15. Estimate the product of 89 and 32.
16. The area of a square is 100 cm^2. What is the measure of each side?
17. $\frac{5}{6} \div \frac{5}{2} = ?$
18. Which digit is in the hundredths place in 24.67?
19. Which factors of 18 are also factors of 24?
20. Solve the proportion for *x*. $\frac{5}{7} = \frac{x}{105}$

1.	2.	3.	4.
5.	6.	7.	8.
9.	10.	11.	12.
13.	14.	15.	16.
17.	18.	19.	20.

Lesson #102

1. $56.7 - 25.34 = ?$

2. $1.2 \div 0.12 = ?$

3. How many yards are in a mile?

4. Round 24,786,098 to the nearest thousand.

5. Find the mean of 93, 99, 93, 87, and 98.

6. Find $\frac{1}{3}$ of 24.

7. $\frac{4}{5} \times \frac{15}{16} = ?$

8. $15\frac{8}{9} + 24\frac{4}{9} = ?$

9. Define *proportion.*

10. Write the decimal number for *fifteen and three hundredths.*

11. A soccer field has a perimeter of 150 yards. If the length is twice as long as the width, what are the dimensions of the field?

12. Find the LCM of 12 and 18.

13. Arrange these decimals in order from greatest to least.

    0.3 0.31 0.301

14. $2\frac{1}{4} \div 4\frac{1}{2} = ?$

15. $\frac{5}{7} \bigcirc \frac{9}{11}$

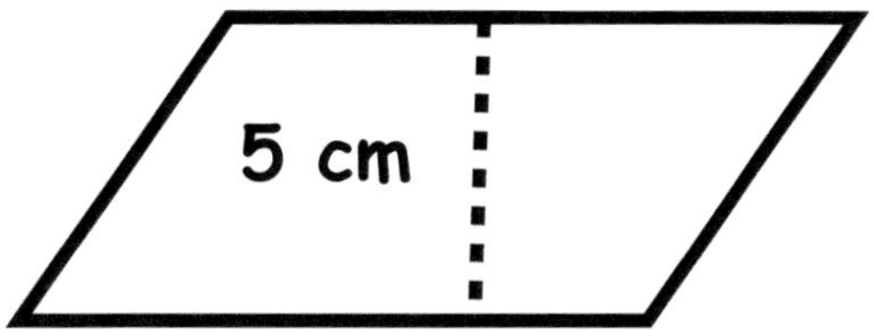

16. Find the area of the parallelogram. A = bh (See *Help Pages.*)

17. $346 + 899 = ?$

18. A triangle with two congruent sides is a(n) _____ triangle.

19. Water boils at _____ degrees Fahrenheit.

20. How many inches are in 3 yards?

1.	2.	3.	4.
5.	6.	7.	8.
9.	10.	11.	12.
13.	14.	15.	16.
17.	18.	19.	20.

Lesson #103

1. A triangle with all sides congruent is a(n) _______ triangle.
2. Write $\frac{3}{5}$ as a decimal and as a percent.
3. $432 \div 30 = ?$
4. Find the area of the parallelogram.

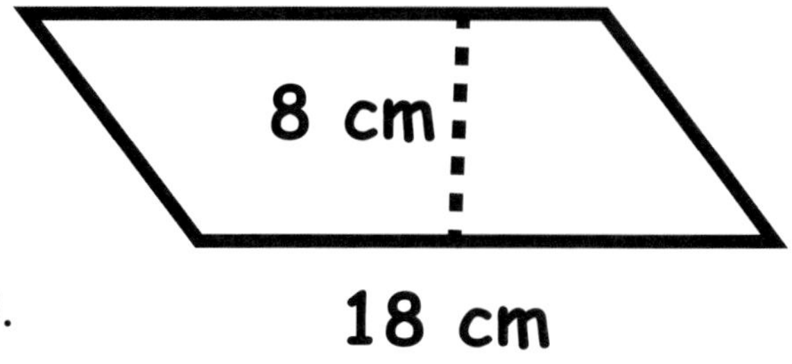

5. $\frac{6}{7} \times \frac{14}{20} = ?$
6. Find the median of 34, 88, 76, 45, and 33.
7. Which digit is in the hundredths place in 9.07?
8. Estimate the product of 497 and 322.
9. The area of a square is 144 mm^2. What is the length of each side?
10. In a class of 30 students $\frac{2}{5}$ are boys. How many students are boys?
11. Solve the proportion for x. $\frac{8}{3} = \frac{x}{120}$
12. If it is 5:30 now, what time will it be in 4 hours and 15 minutes?
13. How many centimeters are in 7 meters?
14. List the factors of 18.
15. Write the formula for finding the volume of a rectangular prism.
16. $2,000 - 799 = ?$
17. A straight angle has how many degrees?
18. How many years are in 8 centuries?
19. $0.67 \times 1.5 = ?$
20. Use the graph to answer the questions.

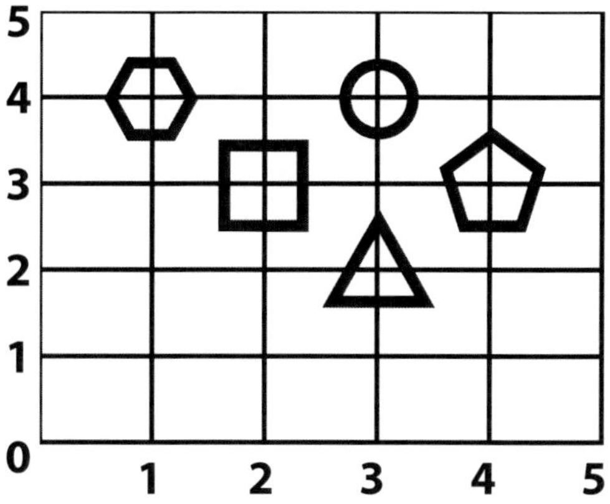

    A) Name the shape located at (3,2).
    B) Name the shape located at (4,3).
    C) Give the pair of numbers that locates the hexagon.

1.	2.	3.	4.
5.	6.	7.	8.
9.	10.	11.	12.
13.	14.	15.	16.
17.	18.	19.	20.

Lesson #104

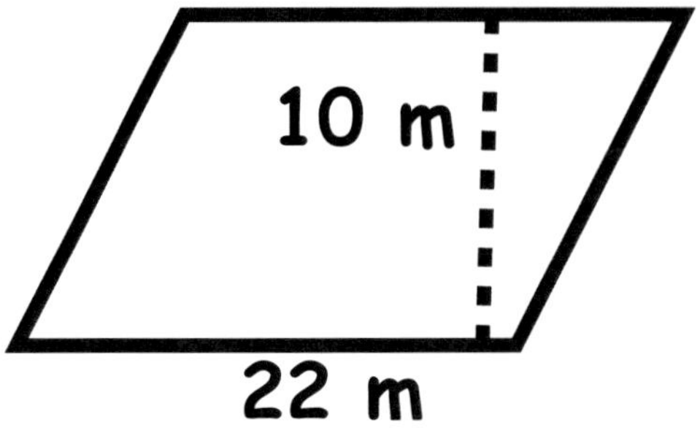

1. Find the area of the parallelogram.
2. $246{,}998 + 655{,}287 = ?$
3. Find $\frac{1}{5}$ of 35.
4. $3\frac{1}{4} - 1\frac{5}{8} = ?$
5. Make a factor tree for 81.
6. Round 246.87 to the nearest tenth.
7. Solve for x. $\frac{7}{8} = \frac{x}{120}$
8. How many ounces are in 8 pounds?
9. The answer to a multiplication problem is the ______.
10. Write the ratio *4:7* in two other ways.
11. Draw a ray.
12. A five-sided shape is called a(n) ________.
13. Find the value of x. $\frac{5}{6} = \frac{x}{36}$
14. $6\frac{1}{4} + 9\frac{2}{5} = ?$
15. Which is greater, 0.865 or 0.65?
16. Write $\frac{14}{5}$ as a mixed number.
17. On the Celsius scale, water boils at ______.
18. $234 \times 121 = ?$
19. Rob is in school 6 hours each day. What fraction of the day is Rob in school?
20. Which letter names the point (1, 4)?
    Which letter names the point (4, 3)?
    Which pair of numbers locates point D?

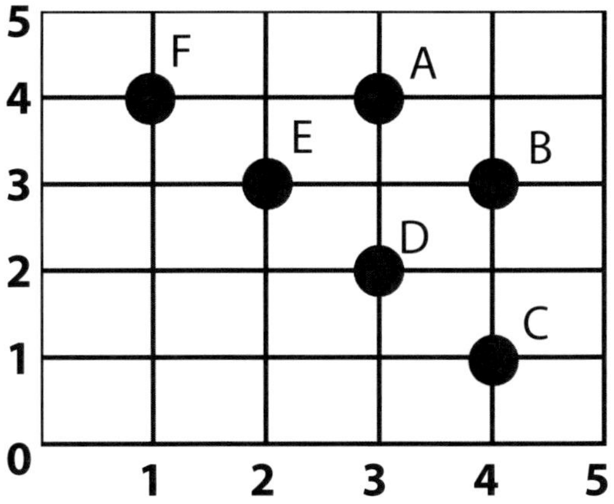

1.	2.	3.	4.
5.	6.	7.	8.
9.	10.	11.	12.
13.	14.	15.	16.
17.	18.	19.	20.

Lesson #105

1. Draw perpendicular lines.
2. Find $\frac{2}{5}$ of 20.
3. $34.56 + 89.9 = ?$
4. $7 - 3\frac{2}{3} = ?$
5. Triangles with no congruent sides are called __________.
6. How many years are in 7 decades?
7. $4.5 \times 6.7 = ?$
8. $\frac{8}{9} \times \frac{9}{14} = ?$
9. Write 0.16 as a percent and as a reduced fraction.
10. $2,907 - 893 = ?$
11. Find the value of x. $\frac{5}{6} = \frac{x}{90}$
12. The area of a square is 64 ft^2. What is the length of each side?
13. How many teaspoons are in 2 tablespoons?
14. It is 2:00 now. What time was it 8 hours ago?
15. The largest cake ever baked was made from 31,026 boxes of cake mix. If each box of cake mix called for 2 eggs, how many dozen eggs were used to make the cake?
16. $83.7 \div 0.9 = ?$
17. Find the LCM of 9 and 15.
18. Write the formula for finding the area of a parallelogram.
19. Would the diameter of a diamond ring best be measured in millimeters or kilometers?
20. If the radius of a circle is 15 feet, what is the diameter?

1.	2.	3.	4.
5.	6.	7.	8.
9.	10.	11.	12.
13.	14.	15.	16.
17.	18.	19.	20.

Lesson #106

1. $47 \times 86 = ?$

2. Round 35,688,900 to the nearest thousand.

3. Solve the proportion for x. $\frac{2}{9} = \frac{x}{81}$

4. $709.8 \div 0.6 = ?$

5. On the Fahrenheit temperature scale, water freezes at ________.

6. Which is greater, 1.009 or 1.09?

7. Fourteen liters is how many milliliters?

8. What do you call the distance around the outside of a circle?

9. Find $\frac{2}{3}$ of 15.

10. Find the area of the parallelogram.

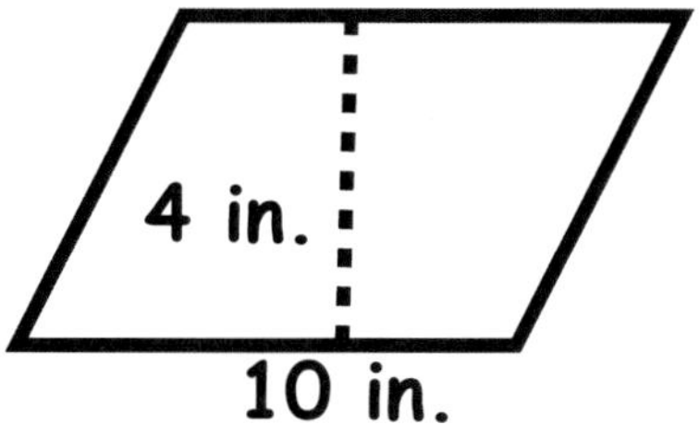

11. $346,887 + 665,219 = ?$

12. Which digit is in the tenths place in 2.78?

13. List the first four prime numbers.

14. Find the mode of 23, 67, 99, 67, and 43.

Find the ordered pair for each building.

15. Hotel ____________________

16. Food Store ________________

17. Bank ____________________

18. Restaurant ________________

19. Train Station ______________

20. Post Office ________________

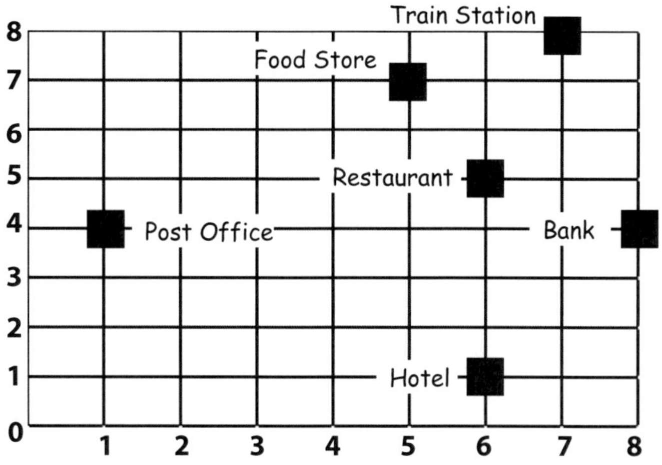

1.	2.	3.	4.
5.	6.	7.	8.
9.	10.	11.	12.
13.	14.	15.	16.
17.	18.	19.	20.

Lesson #107

1. Write 34.56 in words.
2. Write the formula for finding the area of a parallelogram.
3. $4\frac{1}{3} - 2\frac{2}{3} = ?$

4. $\frac{5}{6} + \frac{2}{3} = ?$
5. How many yards are in a mile?
6. 78.9 − 12.76 = ?
7. Which is greater, 0.78 or 0.782?
8. $\frac{4}{5} \bigcirc \frac{8}{10}$
9. List the factors of 35.
10. A triangle with no congruent sides is ______.
11. What is the distance around the outside of a circle called?
12. Put $\frac{6}{12}$ in simplest form.
13. **To find the area of a triangle, multiply the length of the base by the height and divide by 2.** $A = \frac{1}{2}bh$ (See *Help Pages*.)

    Write the formula for finding the area of a triangle.
14. 600 − 98 = ?
15. Find the area of the triangle to the right.

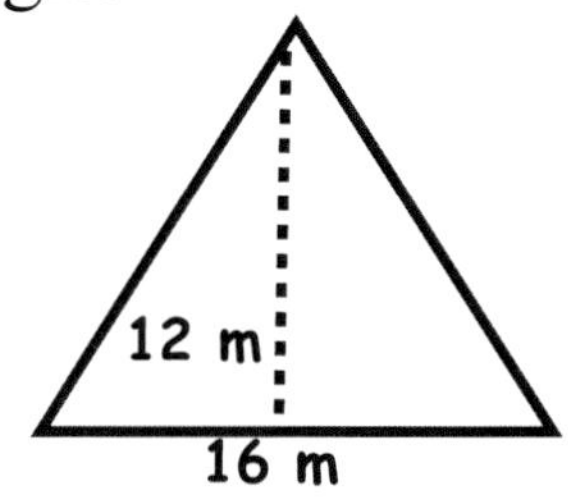

16. How many days are in 9 weeks?
17. An art class has taken orders for 300 ceramic vases. If the students can make 15 vases in a day, how long will it take to fill the order?
18. Is the number 19 a prime or a composite number?
19. Which digit is in the hundred thousands place in 2,568,971?
20. Round 34.826 to the nearest hundredth.

1.	2.	3.	4.
5.	6.	7.	8.
9.	10.	11.	12.
13.	14.	15.	16.
17.	18.	19.	20.

Lesson #108

1. $89.67 + 45.6 = ?$

2. How many ounces are in 3 pounds?

3. $30{,}000 - 18{,}439 = ?$

4. $4\frac{5}{6} + 3\frac{2}{3} = ?$

5. Find the mean of 18, 42, 17, 31, 7, and 23.

6. $6.8 \times 1.7 = ?$

7. Find the area of a square if a side measures 16 centimeters.

8. Find the GCF of 12 and 60.

9. The length of a fishing pole is probably 5 feet or 5 yards?

10. Write the formula for finding the area of a triangle. (Lesson #107)

11. An isosceles triangle has ______ congruent sides.

12. Define *percent.*

13. How many feet are in 4 miles?

14. $\frac{7}{8} - \frac{3}{4} = ?$

15. $\frac{5}{8} \div \frac{3}{8} = ?$

16. Sixty people make up the cast of a local play: 25 are men, 20 are women, and 15 are children. What fraction of the cast are children?

17. Write 42% as a decimal and as a reduced fraction.

18. Estimate the difference of 567 and 235.

19. Find the value of x. $\frac{5}{6} = \frac{x}{96}$

20. Draw a straight angle. How many degrees are in a straight angle?

1.	2.	3.	4.
5.	6.	7.	8.
9.	10.	11.	12.
13.	14.	15.	16.
17.	18.	19.	20.

Lesson #109

1. How many centimeters are in 8 meters?

2. Find the area and the perimeter of a square if a side is 15 cm long.

3. $3.2 \times 0.4 = ?$

4. $\frac{5}{6} - \frac{2}{3} = ?$

5. Find the area of the triangle.

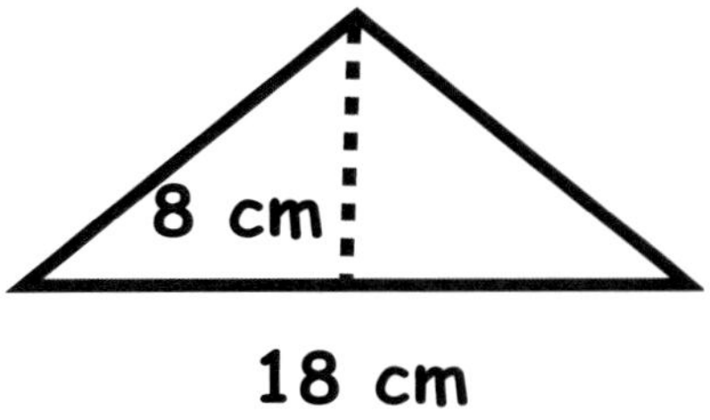

6. Find the LCM of 15 and 25.

7. $40{,}000 - 26{,}788 = ?$

8. What is the value of x? $\frac{3}{5} = \frac{15}{x}$

9. Write the ratio *four to eleven* in two other ways.

10. The California Cougars, a team of 40 players, ate 20 pizzas after their game. How many pizzas would be needed for 120 players?

11. $62 + 19 + 247 = ?$

12. $5 - 2\frac{3}{5} = ?$

13. How many quarts are in 6 gallons?

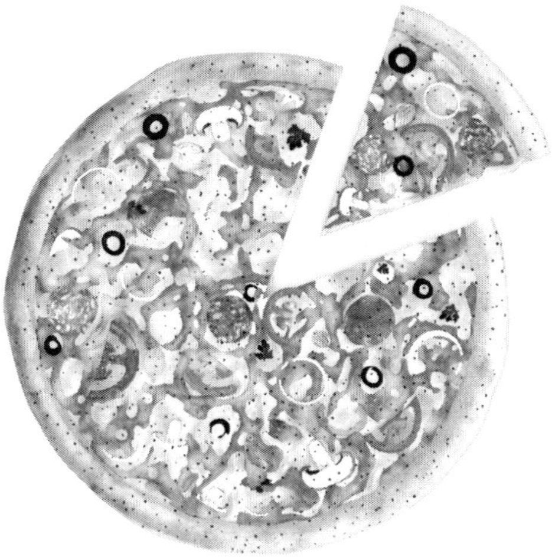

14. Write 0.7623 in words.

15. $55.55 \div 0.05 = ?$

16. $\frac{5}{12} \times \frac{1}{5} = ?$

17. Find $\frac{2}{3}$ of 12.

18. Write $\frac{10}{3}$ as a mixed number.

19. Reduce $\frac{9}{12}$ to simplest terms.

20. Jeff had 60 cookies. How many cookies would he have left, if he gave $\frac{2}{5}$ of them to Mike and $\frac{1}{5}$ of them to Peter?

1.	2.	3.	4.
5.	6.	7.	8.
9.	10.	11.	12.
13.	14.	15.	16.
17.	18.	19.	20.

Lesson #110

1. Find the mode of 32, 37, 32, 35, and 36.

2. Define *obtuse angle.*

3. 1,053 ÷ 27 = ?

4. Write the formula for finding the area of a triangle.

5. 5,516 × 7 = ?

6. Round 36,786,514 to the nearest hundred thousand.

7. Lynn wants to arrive at the baby shower at 2:00 p.m. Her driving time is 1 hour and 20 minutes. She has to stop to buy a card (about 15 minutes) and she plans to stop for lunch (about 45 minutes). At what time should she leave in order to arrive by 2:00?

8. Find the value of x. $\frac{2}{64} = \frac{4}{x}$

9. 3,652 + 45,896 = ?

10. Write the ratio $\frac{5}{7}$ in another way.

11. Put $\frac{12}{15}$ in simplest form.

12. 0.6 − 0.275 = ?

13. 54 ÷ 5 = ?

14. How many degrees make up a straight angle?

15. 0.004 × 0.07 = ?

16. Find $\frac{2}{5}$ of 40.

17. Is 13 a prime or a composite number?

18. Put these decimals in order from least to greatest.

    0.1 0.02 0.356 0.035

19. 0.816 ÷ 6.8 = ?

20. List the factors of 24.

1.	2.	3.	4.
5.	6.	7.	8.
9.	10.	11.	12.
13.	14.	15.	16.
17.	18.	19.	20.

Lesson #111

1. Write $4\frac{5}{8}$ as an improper fraction.
2. Find the area of the parallelogram.

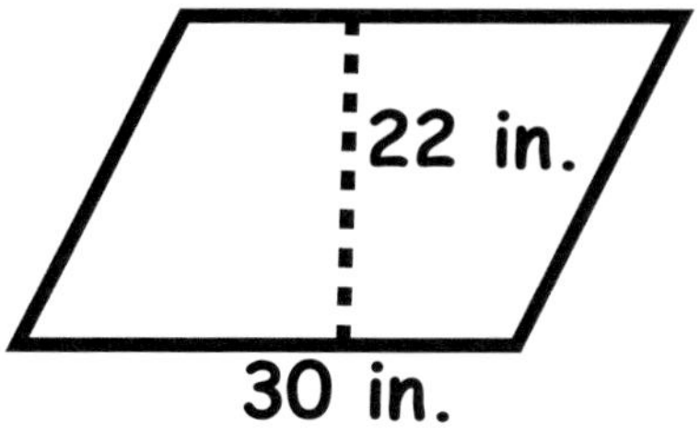

3. $4\frac{3}{4}+2\frac{3}{8}=?$
4. $35 \times 43 = ?$
5. **The formula for finding the circumference of a circle is $C = \pi d$. Multiply the diameter of the circle by the value for *pi* (3.14).** Write the formula.
6. Find the circumference of the circle using the formula above.

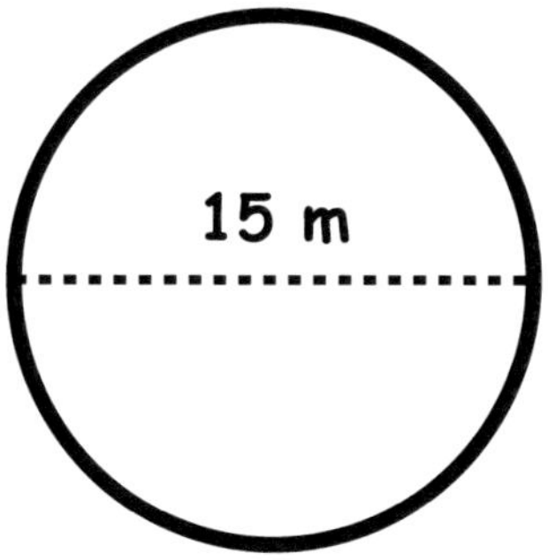

7. $\frac{4}{5}-\frac{1}{4}=?$
8. $\frac{3}{8}\times\frac{12}{15}=?$
9. Draw two congruent rectangles.
10. $5\frac{1}{5}-2\frac{4}{5}=?$
11. The temperature at 9 a.m. was 35°F. At 3 p.m., it was 15° warmer, but by 9 p.m. it had fallen by 25°. Find the temperature at 9 p.m.
12. $1.4 \times 1.4 = ?$
13. Are the ratios $\frac{3}{5}$ and $\frac{9}{15}$ equivalent?
14. $3.069 \div 3.3 = ?$
15. The diameter of a circle is 18 cm. What is the radius?
16. If 5 yards of lace can make a dress, how many yards of lace are needed to make 15 dresses?
17. Find the mean of 310, 421, and 424.
18. Closed figures made up of line segments are________________.
19. If $3x = 15$, what is the value of x?
20. Is this a slide, a rotation or a reflection?

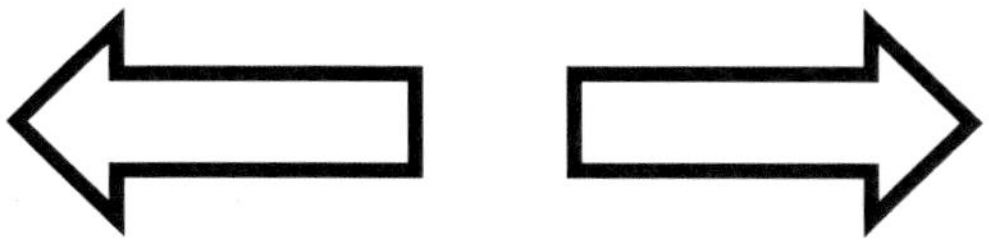

1.	2.	3.	4.
5.	6.	7.	8.
9.	10.	11.	12.
13.	14.	15.	16.
17.	18.	19.	20.

Lesson #112

1. How many years are in 9 decades?
2. Find the GCF of 14 and 21.
3. Find the LCM of 14 and 21.
4. $34.8 + 9.7 + 16.77 = ?$
5. Write 4.753 in words.
6. Round 24,887,663 to the nearest ten thousand.
7. $60{,}000 - 45{,}777 = ?$
8. Round 34.876 to the nearest hundredth.
9. Write the ratio *16:18* in another way and put it in simplest form.
10. Solve the proportion for *a*. $\frac{9}{15} = \frac{a}{10}$
11. What time was it 6 hours and 5 minutes ago, if it is 3:15 now?
12. Find $\frac{3}{5}$ of 25.
13. Find the area of the triangle.

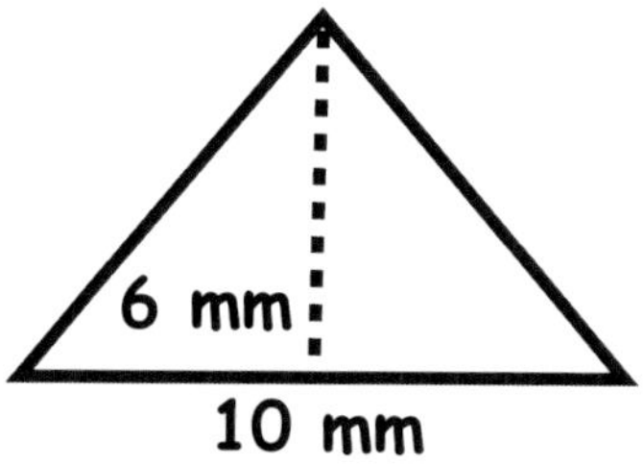

14. Which decimal is smaller, 2.006 or 2.6?
15. Dave drove 2 miles on Tuesday, 4 miles on Wednesday, 8 miles on Thursday, and 16 miles on Friday. If this pattern continues (including the weekend), how far will Dave drive on Monday?
16. $4\frac{1}{2} \times 3\frac{1}{3} = ?$

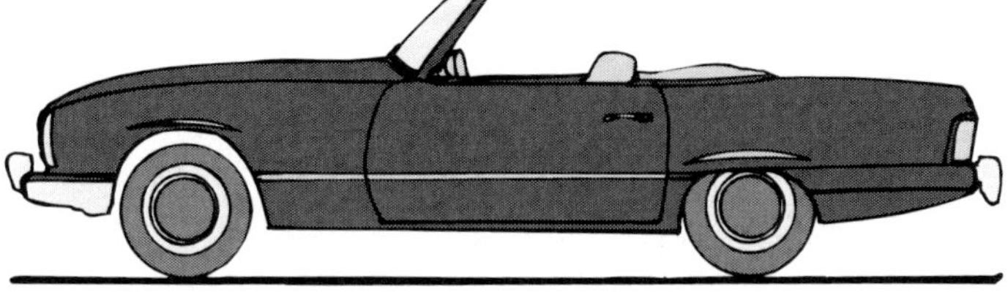

17. $\frac{6}{10} \div \frac{3}{10} = ?$
18. Figures with the same shape, but different sizes are __________.
19. $\frac{5}{7} \bigcirc \frac{3}{5}$
20. Find the circumference of a circle whose diameter is 12 mm.

1.	2.	3.	4.
5.	6.	7.	8.
9.	10.	11.	12.
13.	14.	15.	16.
17.	18.	19.	20.

Lesson #113

1. $1\frac{1}{6} \times 3\frac{1}{7} = ?$
2. 4,579 + 32,886 = ?
3. How many years are in 7 centuries?
4. Find the LCM of 12 and 15.
5. $\frac{4}{10} \div \frac{2}{10} = ?$
6. Write $\frac{34}{5}$ as a mixed number.
7. Write 8.4573 in words.
8. Round 34.87 to the nearest tenth.
9. 50,000 − 29,555 = ?
10. How many inches are in 4 yards?
11. What is the value of x? $\frac{x}{35} = \frac{3}{7}$
12. Make a factor tree for 42.
13. The ratio of soups to salads is eleven to five. Write this ratio in two other ways.
14. Write the formula for finding the area of a triangle.
15. Draw a right angle. How many degrees are in a right angle?
16. Find the circumference of the circle.

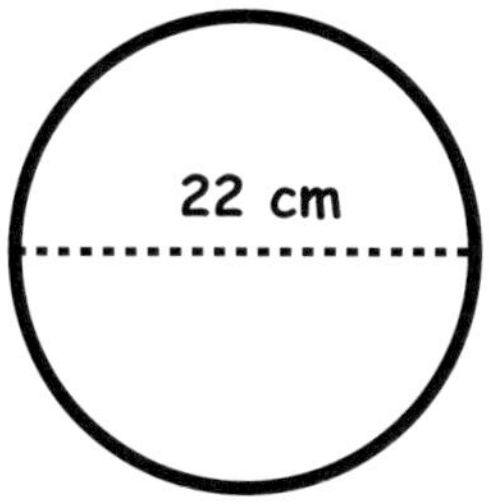

17. $64 \times 35 = ?$
18. $\frac{2}{3} \bigcirc \frac{5}{6}$
19. $\frac{3}{4} + \frac{7}{8} = ?$
20. Find the volume of this cube.

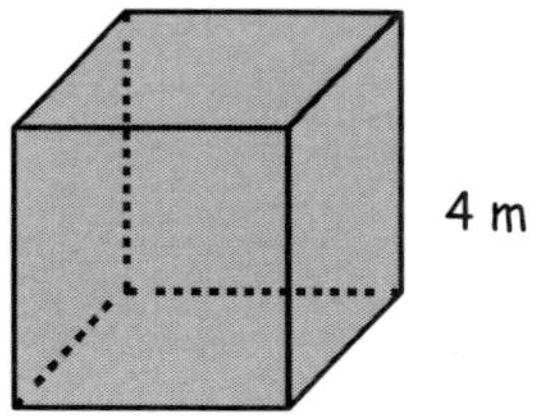

1.	2.	3.	4.
5.	6.	7.	8.
9.	10.	11.	12.
13.	14.	15.	16.
17.	18.	19.	20.

Lesson #114

1. $1.45 \times 0.03 = ?$

2. Put these decimals in order from greatest to least.

   3.46 3.046 3.326 3.406

3. $11\frac{1}{9} - 5\frac{7}{9} = ?$

4. How many quarts are in 7 gallons?

5. Write the ratio *6:14* another way and then put it in simplest form.

6. 5 tablespoons are equal to ______ teaspoons.

7. In a bag of marbles, 3 are blue, 2 are white, 2 are red, and 1 is black. Give the probability of picking a blue one? A black one? A green one?

8. Find the circumference of the circle.

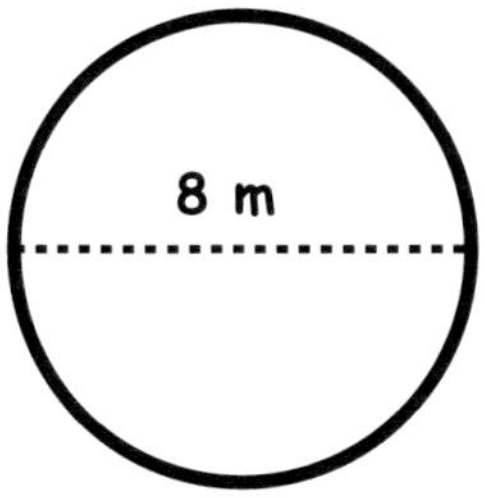

9. $34{,}067 - 21{,}988 = ?$

10. $0.8 - 0.4673 = ?$

11. Round 34,568,992 to the nearest thousand.

12. Find the area of this parallelogram.

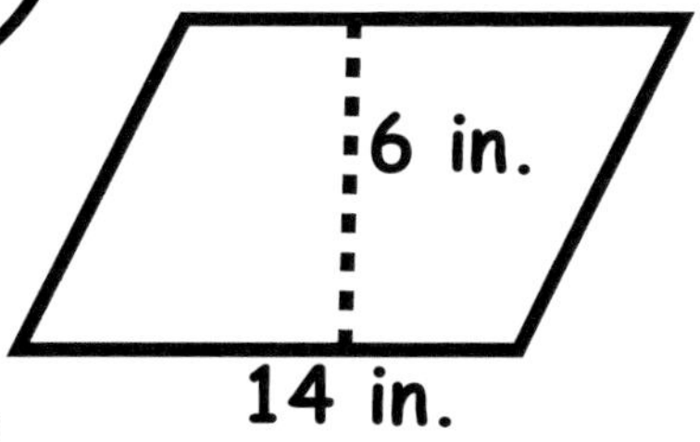

13. Which digit is in the thousandths place in 35.981?

14. $7\frac{2}{3} + 5\frac{1}{4} = ?$

15. $81{,}252 \div 6 = ?$

16. Solve for *n*. $\frac{3}{8} = \frac{9}{n}$

17. There are 12 girls and 15 boys. What is the boy-girl ratio?

18. During which month did the greatest amount of snow fall?

19. About how many inches of snow fell during the whole five months?

20. In which months did 6 in. of snow fall?

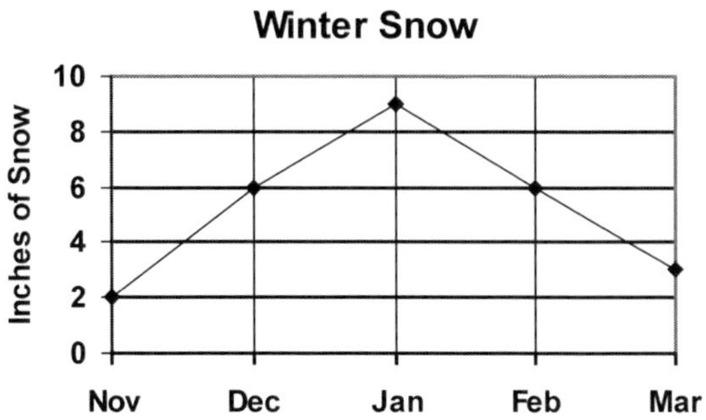

1.	2.	3.	4.
5.	6.	7.	8.
9.	10.	11.	12.
13.	14.	15.	16.
17.	18.	19.	20.

Lesson #115

1. Make a factor tree for 20.
2. If $5x = 30$, what is the value of x?
3. Jess had only $7\frac{1}{2}$ minutes to clean his room. He actually finished cleaning with 45 seconds to spare. How long did it take him to clean his room?

4. $5.62 + 0.8 + 3.0 = ?$
5. $4,568 \div 40 = ?$
6. $0.007 \times 0.009 = ?$
7. $208.4 \div 0.8 = ?$
8. How many cups are in 6 pints?
9. $10\frac{1}{3} + 13\frac{1}{6} = ?$
10. $\frac{5}{7} \times \frac{14}{25} = ?$
11. Find the GCF of 15 and 25.
12. **To find the area of a circle, multiply the radius by itself and multiply that total by *pi* (3.14).** $A = \pi r^2$ Write the formula.

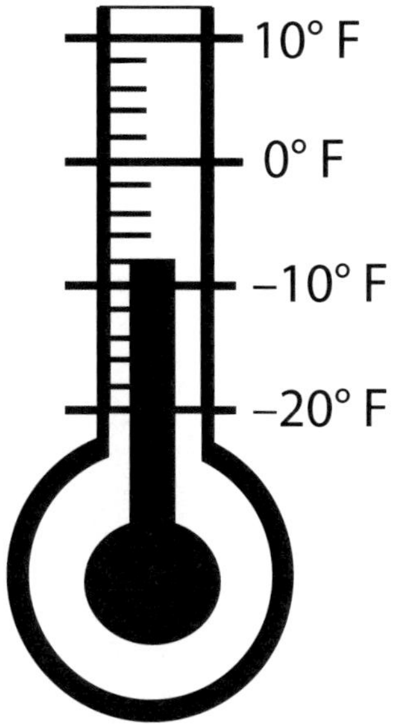

13. $34,702 \times 4 = ?$
14. What temperature is shown on the thermometer?
15. Find the range and the mode of 18, 25, 15, 29, and 18.
16. Is 53 a prime or a composite number?
17. Find the area of the circle.

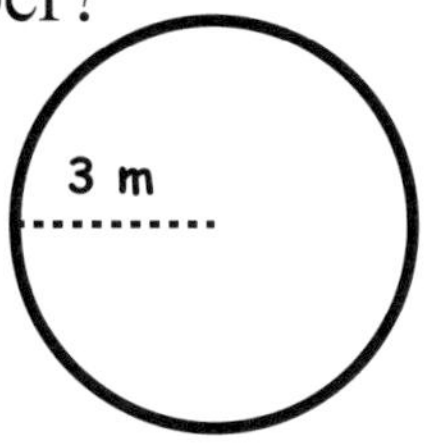

18. Write $\frac{65}{7}$ as a mixed number.
19. $9 - 3\frac{3}{5} = ?$
20. Put these fractions in order from least to greatest. $\frac{2}{3}$ $\frac{1}{6}$ $\frac{2}{9}$

1.	2.	3.	4.
5.	6.	7.	8.
9.	10.	11.	12.
13.	14.	15.	16.
17.	18.	19.	20.

Lesson #116

1. What is the probability of guessing the correct answer to a multiple choice question if the possible answers are *a*, *b*, *c*, and *d*?

2. Find $\frac{4}{5}$ of 20.

3. Find 60% of 50.

4. Find the area of this parallelogram.

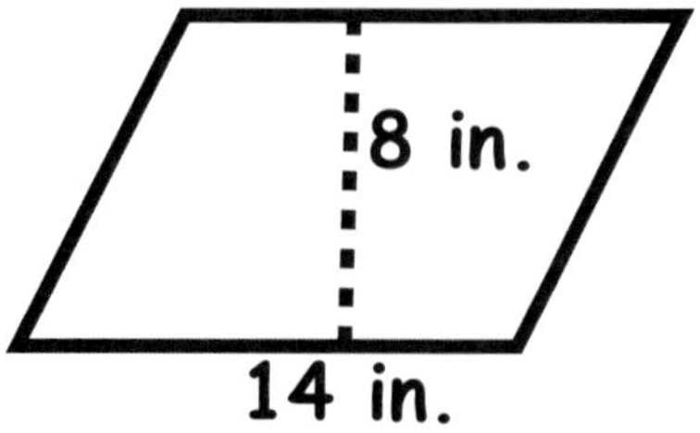

5. What is the name for the number that occurs most often in a list?

6. Write $4\frac{5}{6}$ as an improper fraction.

7. $\frac{4}{9} \bigcirc \frac{7}{10}$

8. Find the LCM of 16 and 20.

9. Find the area of the circle.

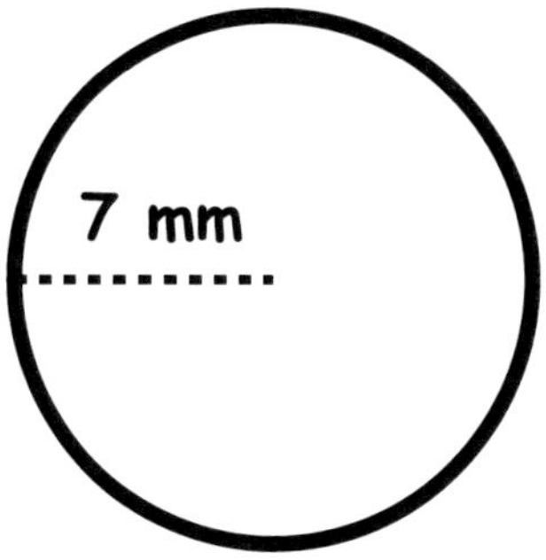

10. $\frac{2}{3} \times \frac{3}{10} = ?$

11. $\frac{48}{8} \div \frac{3}{8} = ?$

12. Water boils at ______ degrees Celsius.

13. On the Fahrenheit scale, water boils at __________.

14. How many centimeters are in 30 meters?

15. What is the distance around the outside of a circle called?

16. Shapes made up of line segments are called __________.

17. Write $\frac{6}{100}$ as a percent.

For 18-20, write whether each is a slide, a rotation, or a reflection.

18. 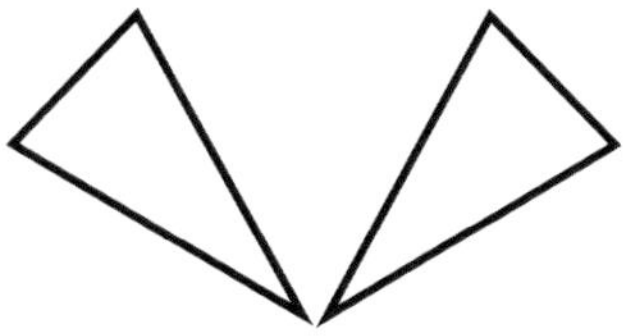

19. 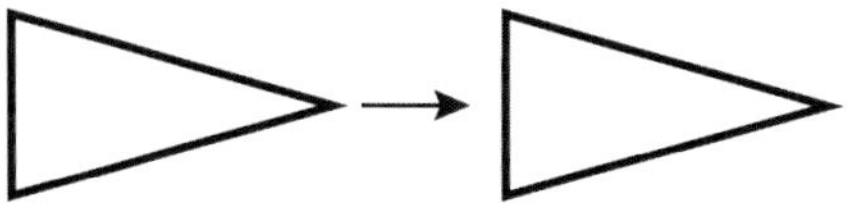

20. 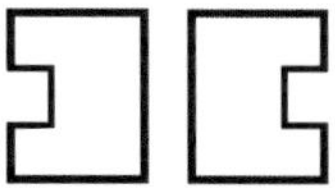

1.	2.	3.	4.
5.	6.	7.	8.
9.	10.	11.	12.
13.	14.	15.	16.
17.	18.	19.	20.

Lesson #117

1. Identify this shape.

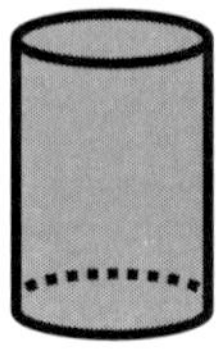

2. $14{,}580 \div 9 = ?$

3. Find the mean of 52, 34, 62, 23, 71, and 52.

4. Find the area of the triangle.

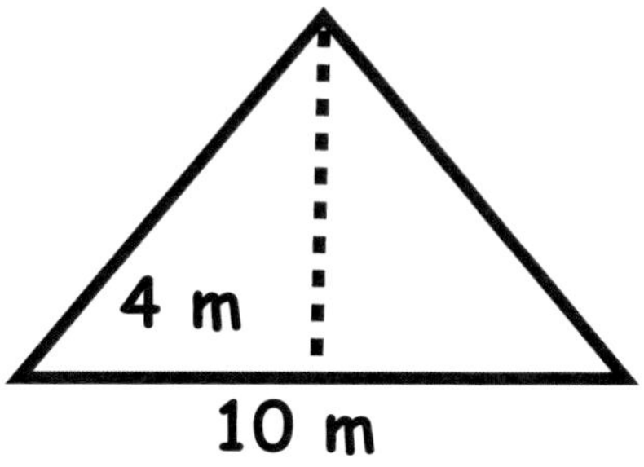

5. Round 32.458 to the nearest hundredth.

6. $34.7 + 9.87 + 0.764 = ?$

7. The area of a square is 36 square inches. What is the length of each side?

8. Give the estimated difference between 653 and 341.

9. Find $\frac{4}{5}$ of 40.

10. Solve to find the value of x. $\frac{x}{4} = \frac{27}{36}$

11. In a bag of marbles, 4 are red, 6 are white and 10 are green. What is the red-white ratio? The white-green ratio? The red-green ratio?

12. Kenny went to the store and bought apples for \$1.30, bread for \$0.99, and 3 cans of pop for \$0.80 each. He was given \$5.35 in change. How much money did he give the cashier?

13. Write $\frac{4}{5}$ as a decimal and as a percent.

14. How many feet are in 7 miles?

15. 0.3 ◯ 0.12

16. $0.48 \div 8 = ?$

17. Write the formula for finding the area of a circle.

18. Find the LCM of 6, 8, and 12.

19. $11\frac{1}{5} - 8\frac{7}{10} = ?$

20. Find the volume of this prism.

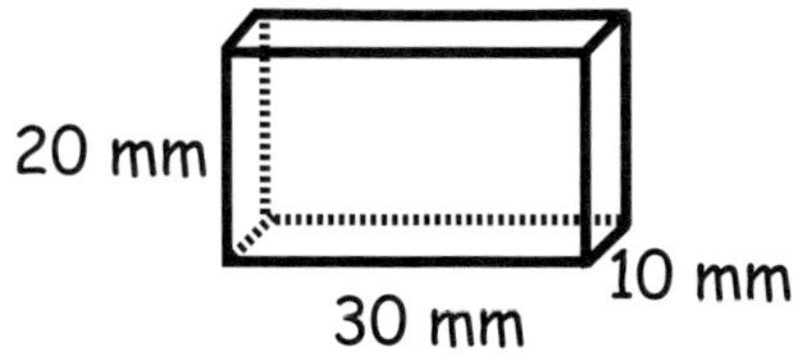

1.	2.	3.	4.
5.	6.	7.	8.
9.	10.	11.	12.
13.	14.	15.	16.
17.	18.	19.	20.

Lesson #118

1. How many pounds are in $7\frac{1}{2}$ tons?
2. $\frac{6}{7} \bigcirc \frac{9}{11}$
3. Find $\frac{2}{3}$ of 36.
4. $\frac{4}{5}+\frac{1}{4}=?$
5. If there were 16 sparrows and 12 bluebirds in the aviary, what was the sparrow-bluebird ratio?
6. Find 40% of 60.
7. Write 0.8 as a fraction and as a percent.
8. Find the difference of $\frac{3}{4}$ and $\frac{2}{5}$.
9. Find the circumference of the circle.

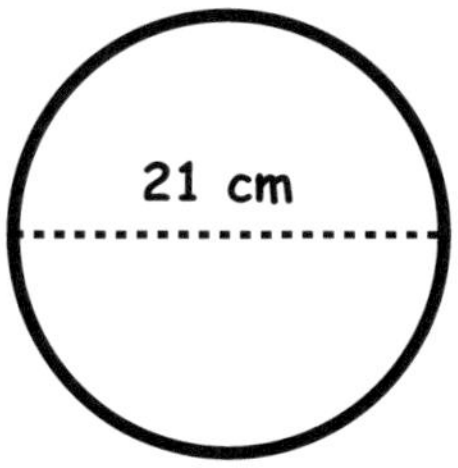

10. Find the perimeter of an equilateral triangle if a side is 25 mm.
11. $2 \div 3\frac{1}{2}=?$
12. If $6x = 24$, what is the value of x?
13. Solve for a. $\frac{12}{15}=\frac{a}{20}$

14. $34.56 + 2.789 + 23.9 = ?$
15. Tina made 4 dozen cookies and ate $\frac{1}{6}$ of them. How many are left?
16. How many millimeters are 4 meters?
17. If it is 4:45 now, what time will it be in 6 hours and 10 minutes?
18. Don is 6 feet 3 inches tall. What is Don's height in inches?
19. Is 41 a prime or a composite number?
20. $0.14 \times 0.16 = ?$

1.	2.	3.	4.
5.	6.	7.	8.
9.	10.	11.	12.
13.	14.	15.	16.
17.	18.	19.	20.

Lesson #119

1. Write the formula for finding the circumference of a circle.

2. Numbers that have only two factors are _______ numbers.

3. $(44 + 45) - (20 \times 3) = ?$

4. $4\frac{1}{6} - 1\frac{5}{6} = ?$

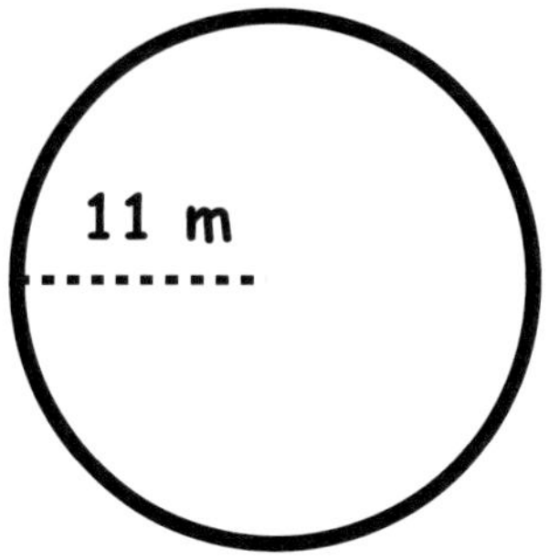

5. Find the area of the circle.

6. $\frac{5}{6} \times \frac{2}{3} = ?$

7. Round 35.763 to the nearest tenth.

8. Find the mean of 2.4, 5.7, and 6.3.

9. Write 9.3121 in words.

10. If you sleep 8 hours a day, what fraction of the day do you sleep?

11. There are four towers on Fisher Street. Tower 1 has three times as many windows as Tower 2. Tower 2 has 30 fewer windows than Tower 3. Tower 3 has 20 more windows than Tower 4. If Tower 4 has 50 windows, how many windows do the others have in all?

12. Find $\frac{5}{6}$ of 24.

13. What is the sum of the first five even numbers?

14. $456 - ? = 279$

15. Find the LCM of 16 and 18.

16. Write the reciprocal of $\frac{3}{4}$.

17. Which digit is in the thousandths place in 6.241?

18. Arrange these decimals in order from greatest to least.

    3.9 3.009 3.87 3.08

19. Make a factor tree for 64.

20. What is the probability of rolling a seven on one throw of a die?

1.	2.	3.	4.
5.	6.	7.	8.
9.	10.	11.	12.
13.	14.	15.	16.
17.	18.	19.	20.

Lesson #120

1. How many minutes are in 7 hours?
2. Write $\frac{3}{10}$ as a decimal and a percent.
3. A triangle with no sides congruent is labeled ___________.
4. $0.54 \div 0.06 = ?$
5. $3.45 \times 0.7 = ?$
6. $\frac{9}{10} - \frac{1}{10} = ?$
7. Find the GCF of 16 and 30.
8. Write $\frac{36}{7}$ as a mixed number.
9. How many years are in 5 decades?
10. What is 70% of 30?
11. Find the mean of 42, 90, and 87.
12. Find the circumference of this circle.
13. $244 \div 6 = ?$

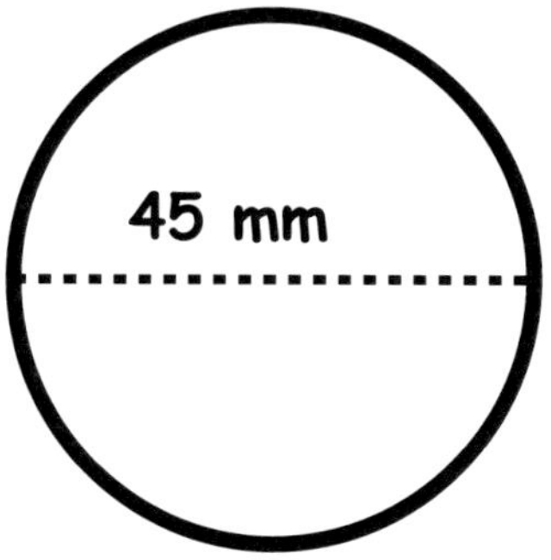

14. What is the probability of an event that is certain to happen?
15. A triangle with two sides congruent is called ___________.
16. Find the volume of a rectangular prism whose length is 5 feet, whose width is 3 feet, and whose height is 3 feet.
17. Draw a hexagon. Show a line of symmetry.
18. How much money do most students receive?
19. How many students earn \$6 per week?
20. How many students earn less than \$5 each week?

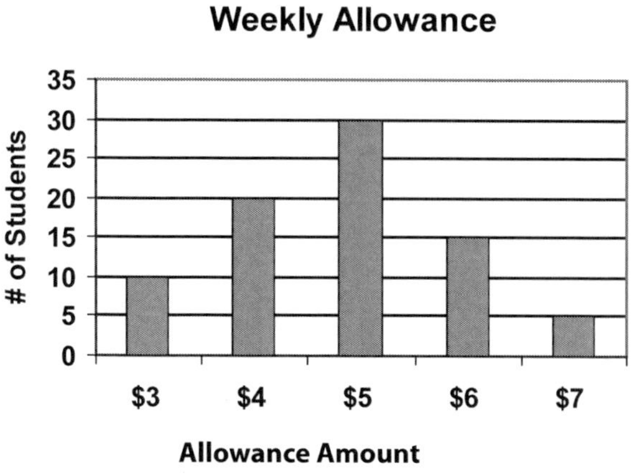

1.	2.	3.	4.
5.	6.	7.	8.
9.	10.	11.	12.
13.	14.	15.	16.
17.	18.	19.	20.

Lesson #121

1. How many centimeters are in 10 meters?
2. Find the median and the mode of 63, 47, 19, 86, and 19.
3. Closed figures made up of line segments are _________.
4. Define *radius*.
5. In Spanish club, the boy-girl ratio was 12 to 15. If 30 girls were members, how many boys were there?
6. **To find the area of a trapezoid, add the top and bottom base, divide that total by 2 and multiply by the height.**
   $A = \left(\frac{\textbf{base + base}}{\textbf{2}} \times \textbf{height}\right)$ Write the formula.
7. Find the area of this trapezoid.

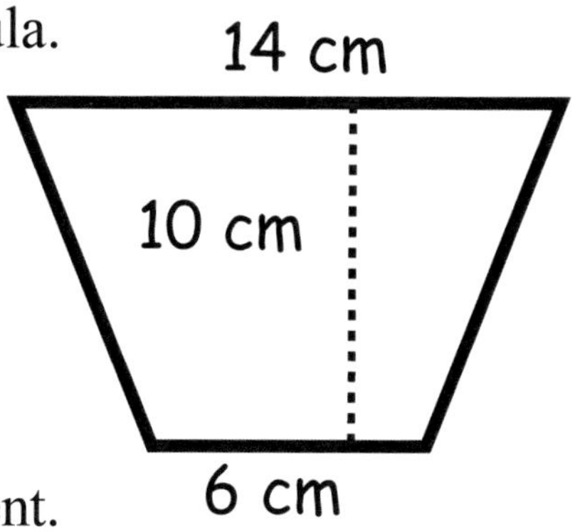

8. $0.072 \times 1.6 = ?$
9. $9\frac{5}{10} + 3\frac{2}{5} = ?$
10. Write 0.6 as a reduced fraction and as a percent.
11. Write the ratio *five to nine* in two other ways.
12. $8{,}642 + 36{,}493 = ?$
13. A number with only two factors is a _________ number.
14. Find $\frac{3}{5}$ of 15.
15. What is 60% of 80?
16. $7.2 - 3.967 = ?$
17. Write 5.4273 using words.
18. What percent of 50 is 15?
19. Round 46,388,910 to the nearest thousand.
20. $800 - 656 = ?$

1.	2.	3.	4.
5.	6.	7.	8.
9.	10.	11.	12.
13.	14.	15.	16.
17.	18.	19.	20.

Lesson #122

1. Find the LCM of 12 and 18.
2. How many ounces are in 5 pounds?
3. A triangle with all sides congruent is called __________.
4. $3\frac{5}{6}+2\frac{2}{3}=?$
5. Find the area of the trapezoid.

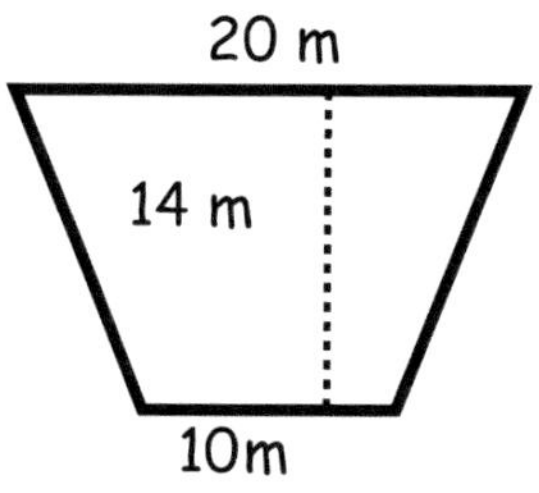

6. $\frac{9}{10}\div\frac{3}{10}=?$
7. Which digit is in the thousandths place in 7.2654?
8. Write the formula for finding the area of a parallelogram.
9. The answer in a subtraction problem is called the __________.
10. 97.65 + 34.9 = ?
11. Estimate the product of 86 and 73.
12. How many nickels are in $3?
13. $\frac{6}{10}\times\frac{5}{9}=?$
14. Find the area of this triangle.

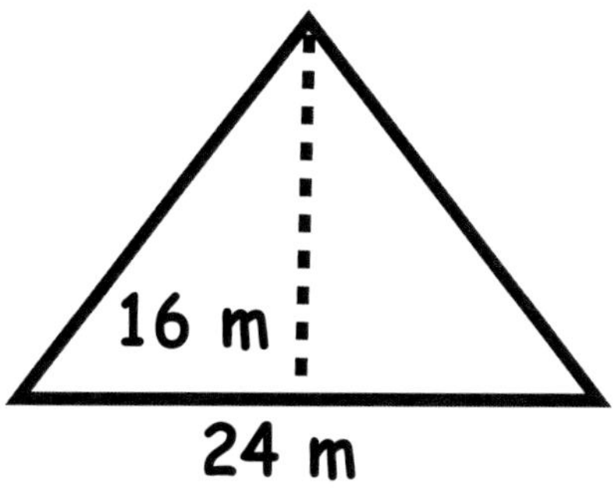

15. Figures with the same size and shape are ___________.
16. How many inches are in 4 feet?
17. Find the GCF of 12 and 20.
18. Make a factor tree for 36.
19. Find the area of a square if one side measures 9 feet.
20. Measure the length.

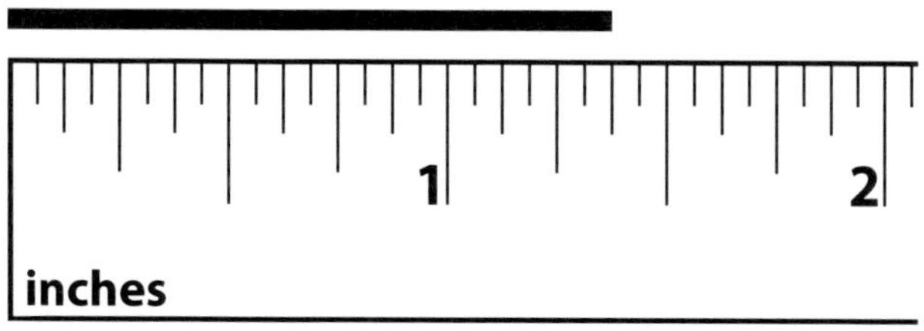

1.	2.	3.	4.
5.	6.	7.	8.
9.	10.	11.	12.
13.	14.	15.	16.
17.	18.	19.	20.

Lesson #123

1. The plans for a shed were drawn so that 1 inch = 3 feet. On the plans, the shed is 5 inches tall, how tall will the real shed be?

2. What is the area of a room that is 12 feet long by 10 feet wide?

3. What is 70% of 50?

4. $\frac{5}{8} \bigcirc \frac{9}{10}$

5. Write $\frac{2}{5}$ as a decimal and as a percent.

6. Solve the proportion for x. $\frac{6}{10} = \frac{9}{x}$

7. Find 15% of 20.

8. How many quarts are in 8 gallons?

9. $1\frac{2}{5} + 4\frac{2}{3} = ?$

10. What kind of angle measures less than 90°?

11. Round 6.742 to the nearest tenth.

12. How many tons are 14,000 pounds?

13. List the factors of 20.

14. Put $\frac{15}{25}$ in simplest form.

15. Find the area of the trapezoid.

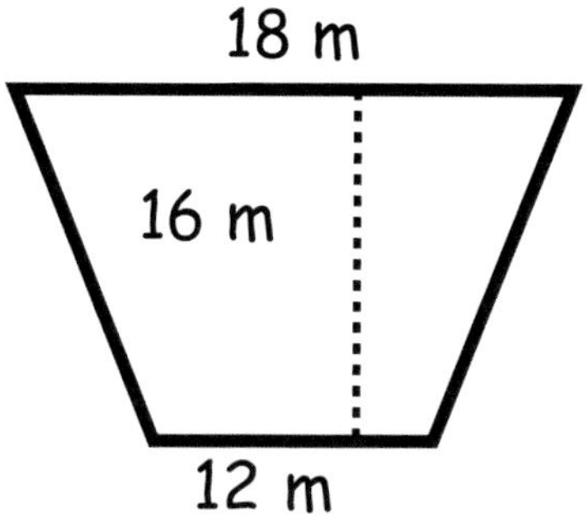

16. $\frac{5}{8} \times \frac{4}{5} = ?$

17. Is the number 23 prime or composite?

18. Find the area of a circle whose radius is 8 mm.

19. Order these decimals from least to greatest. 6.9 6.09 6.19 6.7

20. Bill plans to buy a movie ticket for himself and three of his friends. The cost of a ticket is $5.75. How much will Bill spend on tickets?

1.	2.	3.	4.
5.	6.	7.	8.
9.	10.	11.	12.
13.	14.	15.	16.
17.	18.	19.	20.

Lesson #124

1. $0.4 - 0.276 = ?$

2. $0.525 \div 0.05 = ?$

3. $2 - 1\frac{1}{5} = ?$

4. Find the GCF of 12 and 18.

5. Find the circumference of this circle.

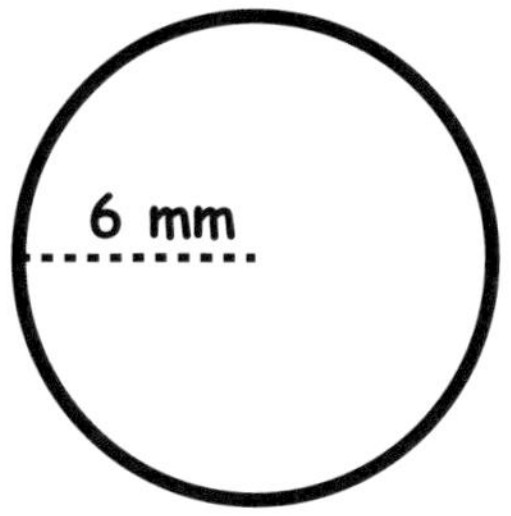

6. What is 20% of 50?

7. How many minutes are in 5 hours?

8. A hot-air balloon flies at a height of 2,561 feet. An airplane is flying 26,784 feet higher than the balloon. How high is the airplane flying?

9. Write 50% as a decimal and as a reduced fraction.

10. Find $\frac{1}{6}$ of 24.

11. Are these figures similar or congruent?

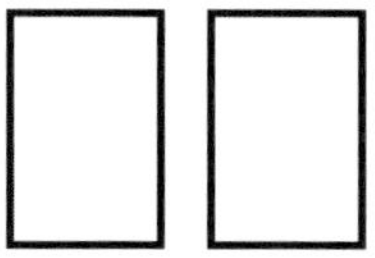

12. Write the formula for finding the area of a circle.

13. Round 23,907,224 to the nearest hundred.

14. What percent of 60 is 18?

15. $45.8 + 9.367 = ?$

16. Put $\frac{8}{10}$ in simplest form.

17. Convert 320 ounces into pounds.

18. How many yards are in a mile?

19. Find the area of the trapezoid.

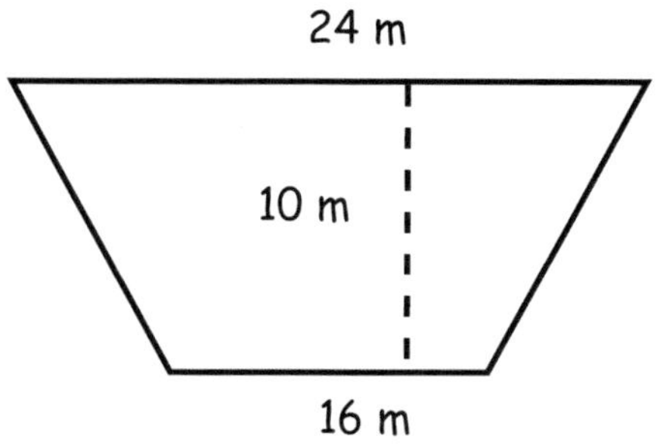

20. In a class of 30 students, there are 14 boys. What is the boy-girl ratio in the class?

1.	2.	3.	4.
5.	6.	7.	8.
9.	10.	11.	12.
13.	14.	15.	16.
17.	18.	19.	20.

Lesson #125

1. How many centimeters are in 4 meters?
2. $\frac{5}{8} \bigcirc \frac{7}{11}$
3. Find $\frac{4}{5}$ of 40.
4. $27 \times 34 = ?$
5. A triangle with two congruent sides is called _______.
6. Write $\frac{34}{6}$ as a mixed number.
7. $0.5 \div 2.5 = ?$
8. $8\frac{1}{4} - 5\frac{3}{4} = ?$
9. Round 3.982 to the nearest hundredth.
10. Make a factor tree for 24.
11. $\frac{7}{10} - \frac{1}{2} = ?$
12. $\frac{8}{10} \div \frac{4}{10} = ?$

13. Write 45% as a decimal and a reduced fraction.
14. 70% of what number is 35?
15. **Surface area is the sum of the areas of all the faces of a solid figure.** ***SA = 2lw + 2wh + 2lh*** Write the formula.
16. Find the surface area of the cube to the right.

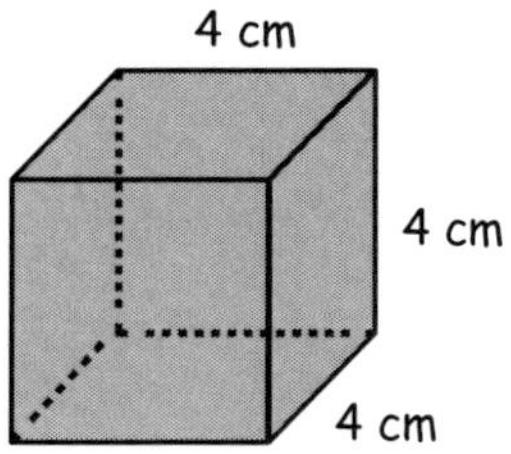

17. Define *percent.*
18. $20.25 \div 0.5 = ?$
19. Give the probability of rolling a 1, 2 or 3 on one roll of a die.
20. State whether the figure was moved by a rotation, a slide, or a reflection.

1.	2.	3.	4.
5.	6.	7.	8.
9.	10.	11.	12.
13.	14.	15.	16.
17.	18.	19.	20.

Lesson #126

1. Draw a cube. How many faces does it have?

2. If it is 5:20 now, what time was it 6 hours and 10 minutes ago?

3. $\frac{5}{6} \times \frac{3}{5} = ?$

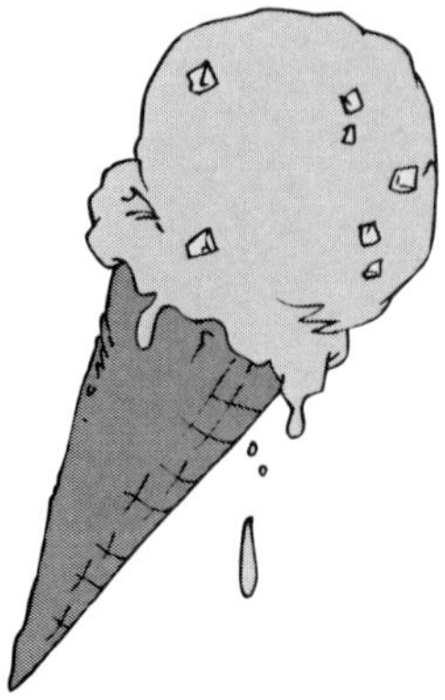

4. $3 - 1\frac{1}{8} = ?$

5. If $7x = 42$, what is the value of x?

6. $0.5 - 0.3682 = ?$

7. How many centuries are 600 years?

8. Write $2\frac{4}{5}$ as an improper fraction.

9. Find the mean of 2.4, 6.3, and 5.7.

10. A single scoop of ice cream weighs 4 ounces. The Dairy Dip used 16 pounds of ice cream yesterday. About how many scoops of ice cream were served yesterday?

11. Find the LCM of 10 and 15.

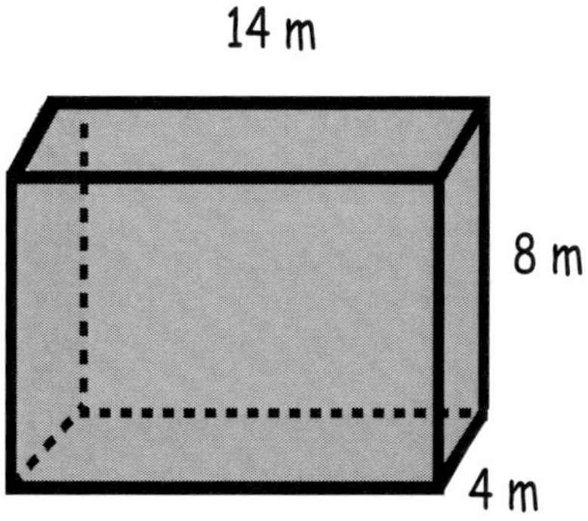

12. $0.009 \times 0.05 = ?$

13. What is the surface area of this solid?

14. $4{,}283 \div 9 = ?$

15. Solve the proportion for x. $\frac{8}{10} = \frac{12}{x}$

16. Find the area of a circle whose radius is 12 cm.

17. Find 40% of 70.

18. 24 is what percent of 80?

19. Find the value of c. $\frac{4}{6} = \frac{12}{c}$

20. $\frac{9}{11} - \frac{3}{11} = ?$

1.	2.	3.	4.
5.	6.	7.	8.
9.	10.	11.	12.
13.	14.	15.	16.
17.	18.	19.	20.

Lesson #127

1. What is the value of x? $\frac{3}{8} = \frac{9}{x}$

2. $17\frac{1}{9} + 22\frac{2}{3} = ?$

3. How many millimeters are 7 meters?

4. Kendra wants to buy a jacket that costs $200. The sales tax is 7%. If the tax is added to the bill, what is the total amount Kendra owes?

5. Write 12% as a decimal and a reduced fraction.

6. Write the ratio $\frac{7}{9}$ in two other ways.

7. Find the GCF of 14 and 21.

8. $7.5 \div 0.5 = ?$

9. $4 - 2\frac{5}{7} = ?$

10. Find the area of the parallelogram.

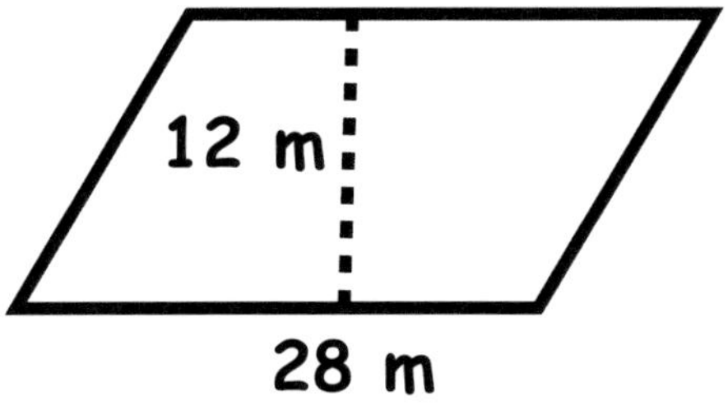

11. $0.9 - 0.8643 = ?$

12. 40% of what number is 28?

13. Draw a ray.

14. The radius of a circle is 17 centimeters. What is the diameter?

15. Find the median and the mode of 56, 77, 89, 32, and 77.

16. Find $\frac{3}{5}$ of 60.

17. The area of a square is 49 ft^2. What is the length of each side?

18. What is 15% of 60?

19. Write the formula for finding the area of a trapezoid.

20. What time will it be in 9 hours and 25 minutes, if it is 7:10 now?

1.	2.	3.	4.
5.	6.	7.	8.
9.	10.	11.	12.
13.	14.	15.	16.
17.	18.	19.	20.

Lesson #128

1. $\frac{5}{12} \bigcirc \frac{7}{11}$

2. Find the estimated sum of 6,788 and 4,321.

3. Make a factor tree for 30.

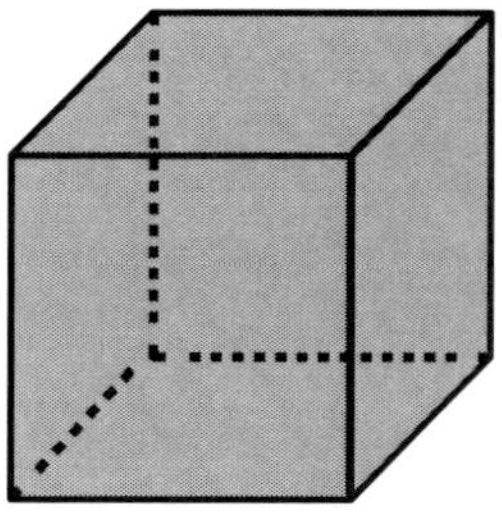

4. $45{,}677 + 98{,}249 = ?$

5. Find the surface area of the cube.

6. $90{,}000 - 34{,}761 = ?$

7. Write the formula for finding the circumference of a circle.

8. $0.089 \times 0.04 = ?$

9. Write 0.12 as a reduced fraction.

10. $0.3 \bigcirc 0.24$

11. How many cups are in 10 pints?

12. $\frac{9}{4} \div 1\frac{1}{2} = ?$

13. $6 - 3\frac{2}{7} = ?$

14. Marcus sleeps 8 hours each night. What fraction of the day does Marcus sleep?

15. List the factors of 24.

16. Round 56.752 to the nearest tenth.

17. Write 9.679 in words.

18. Carl bought 10 bottles of correction fluid at a price of $0.35 each. How much was his bill?

19. Write $\frac{67}{8}$ as a mixed number.

20. $29 \times 35 = ?$

1.	2.	3.	4.
5.	6.	7.	8.
9.	10.	11.	12.
13.	14.	15.	16.
17.	18.	19.	20.

Lesson #129

1. What is 70% of 30?

2. Find the perimeter of the irregular shape.

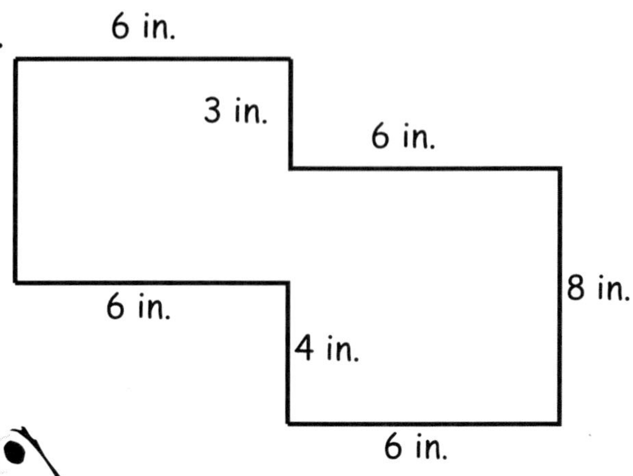

3. $6 - 3\frac{3}{7} = ?$

4. $0.12 \times 0.05 = ?$

5. $82.40 \div 0.04 = ?$

6. If $5x = 35$, what is the value of x?

7. $\frac{9}{12} \bigcirc \frac{8}{11}$

8. Find $\frac{3}{5}$ of 20.

9. $\frac{5}{8} \times \frac{12}{15} = ?$

10. Find the LCM of 12 and 20.

11. How many inches are in 5 yards?

12. **To find the probability of two independent events occurring together, multiply the individual probabilities. [P(A) × P(B)]**

    Example: P(rolling a six on one roll of a die) and P(getting heads on one flip of a coin). $\frac{1}{6} \times \frac{1}{2} = \frac{1}{12}$ Write the formula.

13. Round 34.67 to the nearest tenth.

14. $36 \times 58 = ?$

15. Solve for the value of x. $\frac{9}{12} = \frac{x}{60}$

16. Write 3.736 in words.

17. It is 5:25. What time was it 6 hours and 10 minutes ago?

18. Write 0.55 as a percent and as a reduced fraction.

19. Sue is 5 feet 8 inches tall. How many inches tall is Sue?

20. Find the compound probability of rolling a number greater than 2 on one roll of a die and getting tails on one flip of a coin.

1.	2.	3.	4.
5.	6.	7.	8.
9.	10.	11.	12.
13.	14.	15.	16.
17.	18.	19.	20.

Lesson #130

1. 0.06 ◯ 0.052

2. 863,422 + 36,849 = ?

3. What is the P(a prime number) and P(a number greater than 1) on two rolls of a die?

4. 0.6 − 0.4275 = ?

5. Find the surface area of the prism.

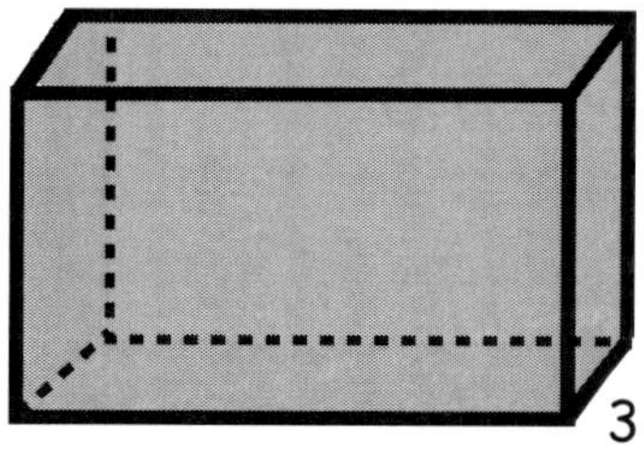

6. Define *mode*.

7. Write $\frac{52}{8}$ as a mixed number.

8. Draw perpendicular lines.

9. Round 42,465,232 to the nearest ten million.

10. What is the name of this shape?

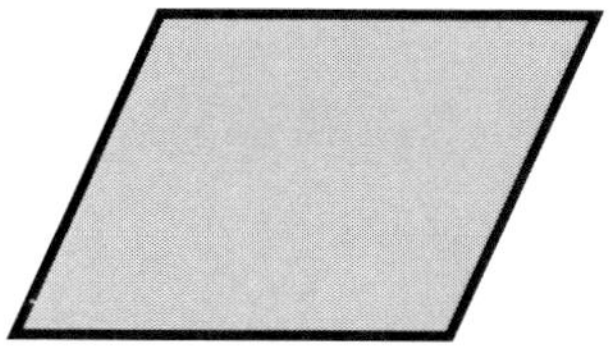

11. Find the value of x. $\frac{12}{15} = \frac{x}{20}$

12. How many years are 12 decades?

13. Write 14% as a decimal and as a reduced fraction.

14. Write $3\frac{3}{5}$ as an improper fraction.

15. 70% of what number is 28?

16. $3.6 \times 0.04 = ?$

17. How many centuries are 500 years?

18. Find the range of 32, 57, 98, 46, and 57.

19. Find the mean of 63, 47, and 70.

20. Sharon had 60 pieces of candy. She gave $\frac{1}{5}$ to her brother and $\frac{3}{5}$ of the candy to her best friend. How many pieces of candy did Sharon have left?

1.	2.	3.	4.
5.	6.	7.	8.
9.	10.	11.	12.
13.	14.	15.	16.
17.	18.	19.	20.

Lesson #131

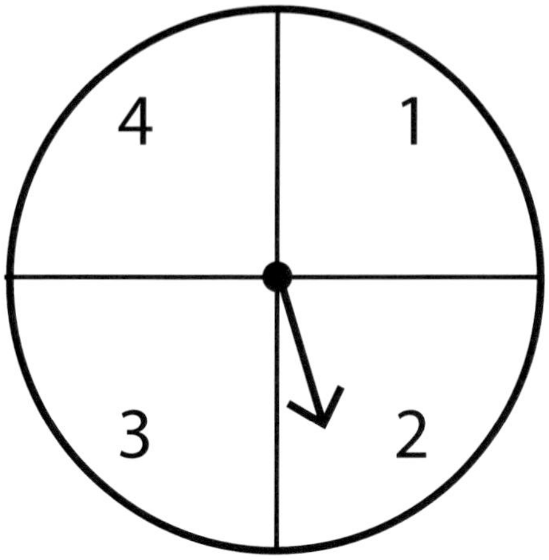

1. Find the GCF and LCM of 8 and 12.
2. What is the value of x? $\frac{5}{6} = \frac{40}{x}$
3. Write $\frac{7}{8}$ as a decimal and as a percent.
4. $36{,}724 \div 4 = ?$
5. How many hours are 2,400 minutes?
6. What is 20% of 50?
7. Find $\frac{2}{3}$ of 90.
8. What is the P(2) and P(an even number) on two spins of this spinner?
9. Figures with the same shape but different sizes are _______.
10. $\frac{6}{10} \times \frac{8}{12} = ?$
11. Draw an acute angle.
12. Find the volume of this rectangular prism.

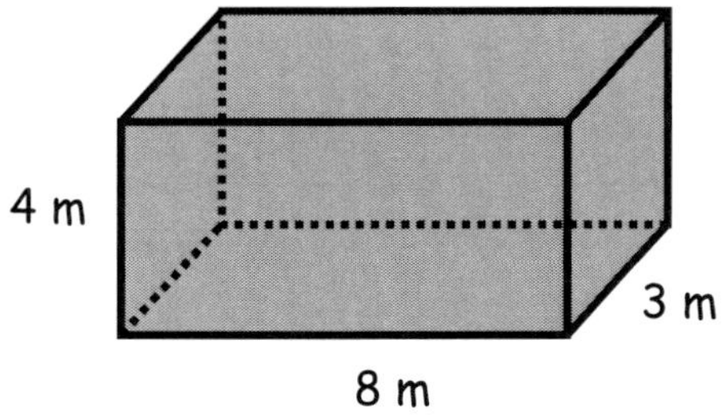

13. Write the formula for finding the area of a parallelogram.
14. How many pounds are $9\frac{1}{2}$ tons?
15. Put these decimals in order from greatest to least.

    3.072 3.7 3.027 3.02

16. Find the circumference of this circle.

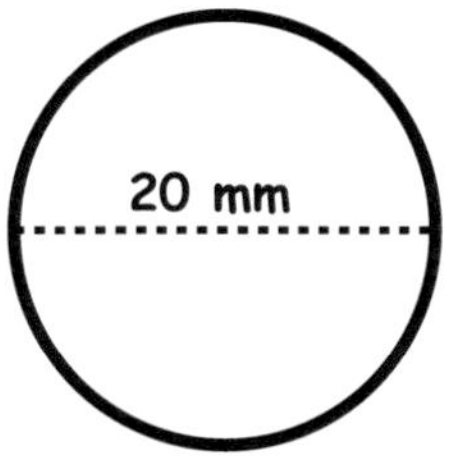

17. List the factors of 16.
18. $0.006 \times 0.004 = ?$
19. If the diameter of a circle is 22 inches, what is the radius?
20. The ratio of boys to girls on the bus is 4 to 3. If there were 16 boys, how many girls are on the bus?

1.	2.	3.	4.
5.	6.	7.	8.
9.	10.	11.	12.
13.	14.	15.	16.
17.	18.	19.	20.

Lesson #132

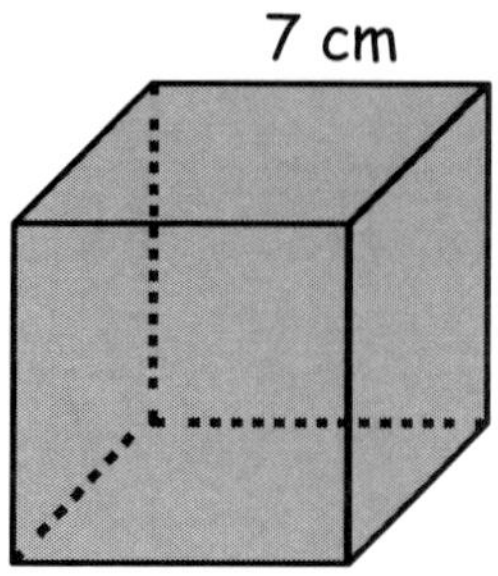

1. Find the surface area of the cube.
2. $4\frac{1}{2} \times 2\frac{1}{3} = ?$
3. Round 37.624 to the nearest hundredth.
4. $\frac{4}{7} \times \frac{14}{16} = ?$
5. Write 50% as a decimal and as a reduced fraction.
6. Write 6.3752 in words.
7. What type of figure is shown here?

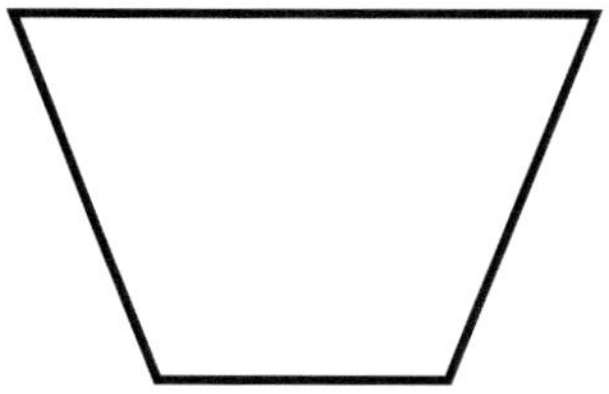

8. Write the ratio *6:11* in two other ways.
9. $\frac{3}{5} \bigcirc \frac{4}{9}$
10. This week, Marquis drove 13 miles on Monday, 17 miles on Tuesday, 21 miles on Wednesday, and 25 miles on Thursday. If this pattern continues, how far will he drive next Wednesday?
11. $\frac{8}{10} \div \frac{4}{10} = ?$
12. $24.86 \div 0.02 = ?$
13. What percent of 80 is 24?
14. What is the P(getting heads on one flip of a coin) and P(rolling a number less than 5 on one roll of a die)?
15. $12\frac{1}{4} - 7\frac{2}{5} = ?$

16. $86.2 + 37.563 = ?$
17. Find the LCM of 6 and 15.
18. How many quarts are in 8 gallons?
19. Closed figures made up of line segments are ____________.
20. On the Celsius temperature scale, water boils at __________.

1.	2.	3.	4.
5.	6.	7.	8.
9.	10.	11.	12.
13.	14.	15.	16.
17.	18.	19.	20.

Lesson #133

1. $9\frac{3}{7} - 6\frac{6}{7} = ?$

2. On the Fahrenheit scale, water freezes at __________.

3. Round 42.654 to the nearest tenth.

4. Solve the proportion for x. $\frac{8}{12} = \frac{6}{x}$

5. Write 0.16 as a fraction and as a percent.

6. Find the mean of 44, 54, and 49.

7. $60{,}000 - 27{,}988 = ?$

8. An eight-sided shape is a(n) ___________.

9. **The set of integers includes the counting numbers, their opposites (negative numbers), and zero.** Put these integers in order from least to greatest. −15 −4 −39 −1 (See *Help Pages*.)

10. $4{,}862 \times 5 = ?$

11. 3.06 ◯ 3.006

12. Numbers that have only two factors are ___________ numbers.

13. Find the GCF of 15 and 25.

14. $0.05 \times 0.04 = ?$

15. Find the P(number greater than 1) and P(2) on two spins of the spinner at the bottom of the page.

16. Write the formula for finding the area of a parallelogram.

17. What will be the time 90 minutes after noon?

18. $8\frac{1}{5} + 4\frac{1}{2} = ?$

19. Find $\frac{2}{5}$ of 40.

20. Write $3\frac{2}{3}$ as an improper fraction.

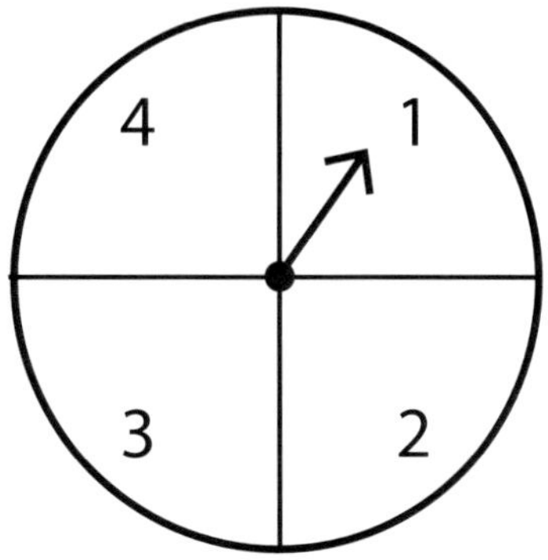

1.	2.	3.	4.
5.	6.	7.	8.
9.	10.	11.	12.
13.	14.	15.	16.
17.	18.	19.	20.

Lesson #134

1. On the Celsius scale, at what temperature does water freeze?

2. List the first five prime numbers.

3. $\frac{8}{10} \times \frac{12}{16} = ?$

4. Put these integers in order from greatest to least.

   −44 0 −16 12 −2 25

5. Write $\frac{34}{6}$ as a mixed number.

6. Find the area of the triangle.

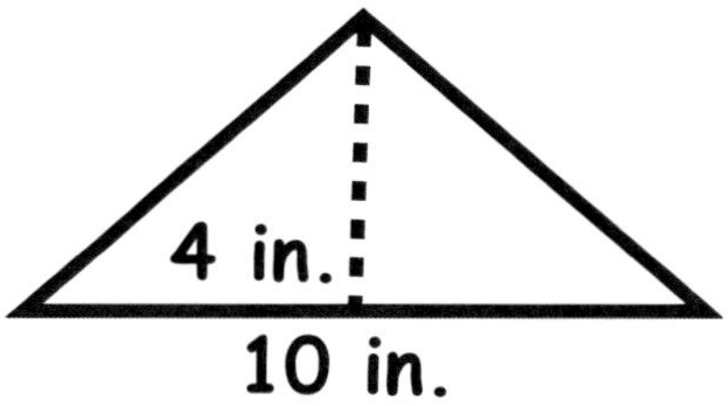

7. Make a factor tree for 50.

8. $0.4 - 0.163 = ?$

9. $12 - 8\frac{4}{5} = ?$

10. List the factors of 24.

11. Find the area of a circle whose diameter is 18 cm.

12. Write the reciprocal of $\frac{5}{8}$.

13. Round 3.786 to the nearest tenth.

14. Write $\frac{7}{10}$ as a decimal and as a percent.

15. $86 + 25 + 47 = ?$

16. How many feet are in 4 miles?

17. Tickets for the school carnival sell for $12 each. If you buy 10 or more, you receive one-fourth off the price of each ticket. Mrs. Sampson wants to buy 14 tickets. How much will her tickets cost?

18. Put $\frac{12}{15}$ in simplest form.

19. Write $\frac{2}{5}$ as a decimal.

20. A triangle with all congruent sides is a(n) __________ triangle.

1.	2.	3.	4.
5.	6.	7.	8.
9.	10.	11.	12.
13.	14.	15.	16.
17.	18.	19.	20.

Lesson #135

1. $\frac{9}{10} - \frac{2}{3} = ?$

2. It is 6:30. What time will it be in 80 minutes?

3. Larry's bedroom has a rectangular shape with an area of 120 ft^2 and a perimeter of 44 feet. What are the dimensions of Larry's room?

4. How many decades are in 50 years?

5. Put these integers in order from least to greatest.

   −11   9   0   −56   −1

6. Write 14.026 in words.

7. Find $\frac{3}{5}$ of 20.

8. 36.7 + 94.82 = ?

9. How many meters are in 600 centimeters?

10. Find 80% of 30.

11. Find the value of x. $\frac{5}{8} = \frac{30}{x}$

12. Write 65% as a reduced fraction.

13. If $9x = 27$, what is the value of x?

14. $3.4 \times 2.6 = ?$

15. Which digit is in the thousandths place in 6.473?

16. Is 46 a prime or a composite number?

17. Round 57.894 to the nearest tenth.

18. Make a factor tree for 28.

19. Find the area of the trapezoid.

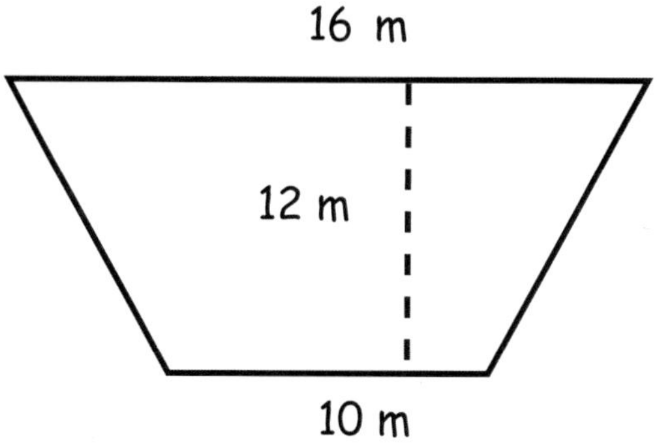

20. Put $\frac{20}{25}$ in simplest form.

1.	2.	3.	4.
5.	6.	7.	8.
9.	10.	11.	12.
13.	14.	15.	16.
17.	18.	19.	20.

Lesson #136

1. $45.05 \div 0.05 = ?$

2. $700 - 463 = ?$

3. Solve for x. $\frac{4}{5} = \frac{x}{60}$

4. Put these integers in order from greatest to least.

   $-66 \quad -18 \quad -6 \quad -89$

5. $1\frac{3}{5} + 4\frac{3}{4} = ?$

6. What is 90% of 40?

7. $7.3 - 5.836 = ?$

8. Which digit is in the thousandths place in 17.375?

9. Write 60% as a decimal and as a reduced fraction.

10. Put 1.6, 1.062, and 1.006 in order from least to greatest.

11. $14\frac{1}{8} - 10\frac{5}{8} = ?$

12. $0.012 \times 0.07 = ?$

13. Find the surface area of this prism.

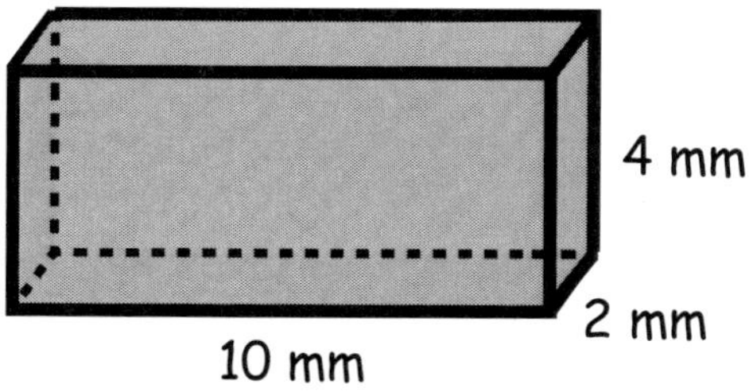

14. Figures with the same size and shape are __________.

15. $\frac{8}{9} \times \frac{3}{16} = ?$

16. $93{,}475 + 46{,}288 = ?$

17. A new mountain bike costs \$300, but you can get a 20% discount if you purchase one today. How much would you save if you bought a bike today?

18. Write 7.62 in words.

19. Write the reciprocal of $\frac{12}{15}$.

20. How many cups are in 7 pints?

1.	2.	3.	4.
5.	6.	7.	8.
9.	10.	11.	12.
13.	14.	15.	16.
17.	18.	19.	20.

Lesson #137

1. A triangle with no congruent sides is a(n) _______ triangle.
2. $3\frac{1}{5}+4\frac{3}{10}=?$
3. How many centimeters are in 15 meters?
4. If $8x = 72$, what is the value of x?
5. The area of a square is 64 square feet. How long is each side?
6. $2.8 - 0.9742 = ?$
7. Find $\frac{4}{5}$ of 25.
8. The boy-girl ratio at the park was 9 to 7. If there were 54 boys, how many girls were at the park?
9. $7-2\frac{3}{5}=?$
10. Write the formula for finding the surface area of a rectangular prism.
11. Jeffery rode 88.24 miles in an 8-hour period. What was the average number of miles he rode in each hour?
12. What is the distance around the outside of a circle called?
13. Write $\frac{3}{10}$ as a decimal and as a percent.
14. $\frac{6}{7} \bigcirc \frac{5}{8}$
15. How many millimeters are 4 meters?
16. $3\div 7\frac{1}{2}=?$
17. Find the area of the parallelogram.

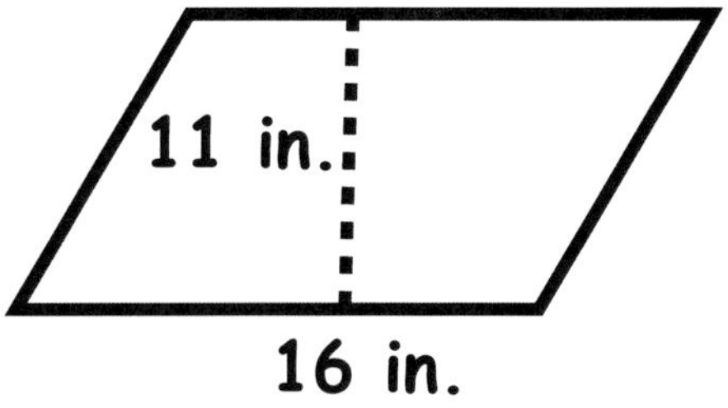

18. What is the P(1, 6, 4) on three rolls of a die?
19. In degrees Fahrenheit, at what temperature does water boil?
20. Make a factor tree for 18.

1.	2.	3.	4.
5.	6.	7.	8.
9.	10.	11.	12.
13.	14.	15.	16.
17.	18.	19.	20.

Lesson #138

1. Write $3\frac{4}{7}$ as an improper fraction.
2. Closed figures made up of line segments are __________.
3. Write $\frac{1}{3}$ as a decimal.
4. What is the 5th prime number?
5. Identify this shape.

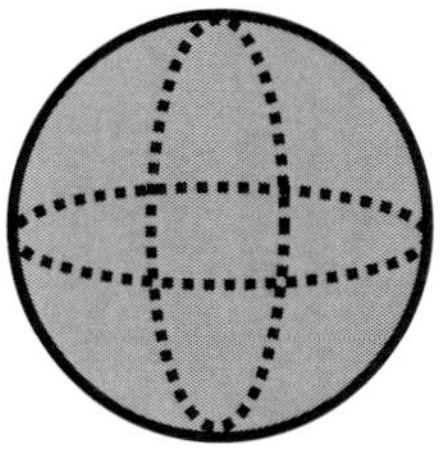

6. A regular octagon has a perimeter of 56 in. How long is each side?
7. Write the formula for finding the area of a triangle.
8. $8\frac{3}{5} - 4\frac{4}{5} = ?$
9. 42 is what percent of 60?
10. Find the circumference of a circle whose diameter is 13 cm.
11. $24 \div 0.08 = ?$
12. What is the median of the numbers 67, 89, 20, 44, and 16?
13. Put these integers in order from least to greatest.

    −56 −22 −2 −76

14. Six of 30 students earned a B on the math test. What fraction of the students got a B on the test? What percent of students got a B?
15. Clay had 75 crayons. He gave $\frac{2}{5}$ of them to his sister and $\frac{1}{5}$ of them to his little brother. How many crayons did Clay have left?
16. Draw a ray.
17. $50{,}000 - 27{,}886 = ?$
18. It is 9:15. What time was it 7 hours and 5 minutes ago?
19. Find the LCM of 14 and 16.
20. $36\frac{2}{5} + 41\frac{2}{3} = ?$

1.	2.	3.	4.
5.	6.	7.	8.
9.	10.	11.	12.
13.	14.	15.	16.
17.	18.	19.	20.

Lesson #139

1. 345,667 + 29,726 = ?

2. What is the P(H, T, T, H) on four flips of a coin?

3. How many inches are in 6 feet?

4. 0.8 − 0.6773 = ?

5. $\frac{5}{8} \times \frac{2}{3} = ?$

6. List the factors of 20.

7. 2,025 ÷ 25 = ?

8. Which digit is in the hundredths place in 5.602?

9. Find the area of a circle whose radius is 8 mm.

10. Put these integers in order from least to greatest.

11. Find the perimeter of a rectangle with a length of 12 meters and a width of 5 meters.

12. Find $\frac{5}{6}$ of 30.

13. 70% of what number is 35?

14. Write 0.15 as a percent and as a reduced fraction.

15. A triangle with two sides congruent is a(n) ___ triangle.

16. If $5n = 45$, what is the value of n?

17. Round 34.744 to the nearest tenth.

18. 0.009 × 0.06 = ?

19. Write the ratio *12:19* in two other ways.

20. Find the surface area of this figure.

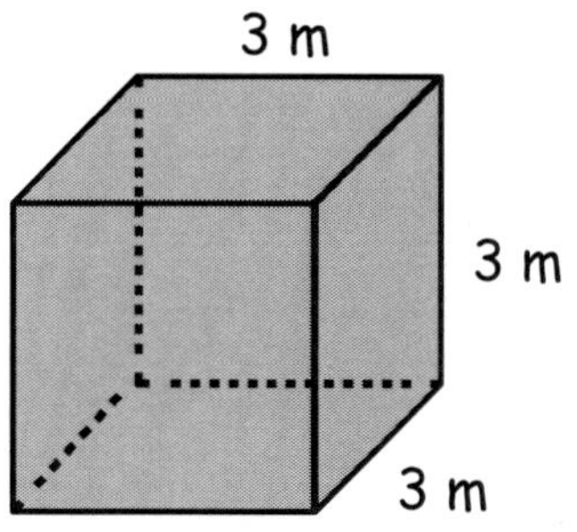

1.	2.	3.	4.
5.	6.	7.	8.
9.	10.	11.	12.
13.	14.	15.	16.
17.	18.	19.	20.

Lesson #140

1. Write the formula for finding the surface area of a rectangular prism.

2. $0.7 - 0.3326 = ?$

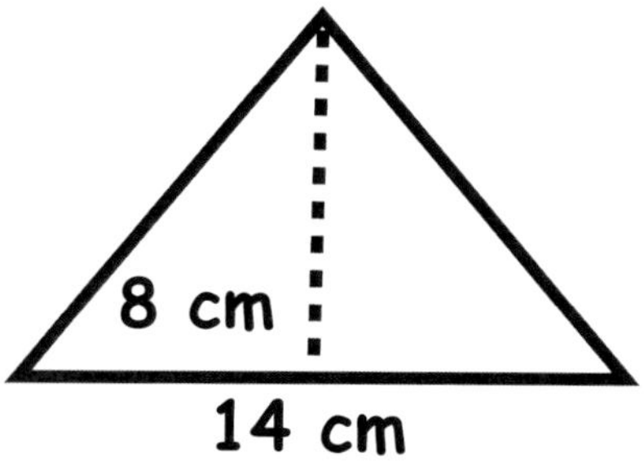

3. $\frac{4}{7} \bigcirc \frac{10}{11}$

4. $\frac{6}{10} \div \frac{2}{10} = ?$

5. Find the area of the triangle.

6. The radius of a circle is 14 mm. What is its diameter?

7. The temperature was 12°F at noon. By 9:00 p.m. it had dropped 15°F. What was the temperature at 9:00 p.m.?

8. How many gallons are 28 quarts?

9. Write $\frac{15}{2}$ as a mixed number.

10. Write $\frac{3}{20}$ as a decimal and as a percent.

11. 90% of what number is 63?

12. Round 496,867,552 to the nearest hundred million.

13. Define *range*.

14. On the Fahrenheit temperature scale, water boils at ___________.

15. If $12x = 36$, what is the value of x?

16. Make a factor tree for 45.

17. How many inches are in 2 yards?

18. $7{,}890{,}654 + 3{,}442{,}897 = ?$

19. What is the probability of drawing a jack out of a deck of 52 cards?

20. $-66 \bigcirc -23$

1.	2.	3.	4.
5.	6.	7.	8.
9.	10.	11.	12.
13.	14.	15.	16.
17.	18.	19.	20.

Simple Solutions.

Minutes a Day–Mastery for a Lifetime!

Intermediate B

Mathematics

Help Pages & "Who Knows?"

Help Pages

Vocabulary

Arithmetic Operations

difference — the result or answer to a subtraction problem. **Example**: The difference of 5 and 1 is 4.

product — the result or answer to a multiplication problem. **Example**: The product of 5 and 3 is 15.

quotient — the result or answer to a division problem. **Example**: The quotient of 8 and 2 is 4.

sum — the result or answer to an addition problem. **Example**: The sum of 5 and 2 is 7.

Factors and Multiples

factors — are multiplied together to get a product. **Example**: 2 and 3 are factors of 6.

multiples — can be evenly divided by a number. **Example**: 5, 10, 15 and 20 are multiples of 5.

composite number — a number with more than 2 factors.

Example: 10 has factors of 1, 2, 5, and 10. Ten is a composite number.

prime number — a number with exactly 2 factors (the number itself and 1). 1 is not prime (it has only 1 factor). **Example**: 7 has factors of 1 and 7. Seven is a prime number.

greatest common factor (GCF) — the highest factor that 2 numbers have in common.

Example: The factors of 6 are 1, 2, 3, and 6. The factors of 9 are 1, 3, and 9. The GCF of 6 and 9 is 3.

least common multiple (LCM) — the smallest multiple that 2 numbers have in common.

Example: Multiples of 3 are 3, 6, 9, 12, 15... Multiples of 4 are 4, 8, 12, 16... The LCM of 3 and 4 is 12.

prime factorization — a number, written as a product of its prime factors.

Example: 140 can be written as $2 \times 2 \times 5 \times 7$ or $2^2 \times 5 \times 7$. (All of these are prime factors of 140.)

Fractions and Decimals

improper fraction — a fraction in which the numerator is larger than the denominator. **Example**: $\frac{9}{4}$

mixed number — the sum of a whole number and a fraction. **Example**: $5\frac{1}{4}$

reciprocal — a fraction where the numerator and denominator are interchanged. The product of a fraction and its reciprocal is always 1.

Example: The reciprocal of $\frac{3}{5}$ is $\frac{5}{3}$. $\frac{3}{5} \times \frac{5}{3} = \frac{15}{15} = 1$

repeating decimal — a decimal in which a number or a series of numbers continues on and on.

Example: 2.33333333, 4.151515151515, 7.125555555, etc.

Geometry

acute angle — an angle measuring less than 90°.

complementary angles — two angles whose measures add up to 90°.

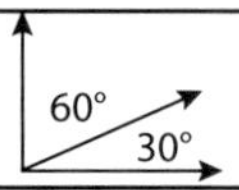

congruent — figures with the same shape and the same size.

obtuse angle — an angle measuring more than 90°.

right angle — an angle measuring exactly 90°.

similar — figures having the same shape, but different size.

Help Pages

Vocabulary (continued)

Geometry
straight angle — an angle measuring exactly 180°.
supplementary angles — two angles whose measures add up to 180°. 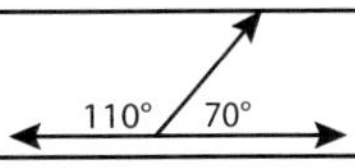
surface area — the sum of the areas of all of the faces of a solid figure.

Geometry — Circles
circumference — the distance around the outside of a circle.
diameter — the widest distance across a circle. The diameter always passes through the center.
radius — the distance from any point on the circle to the center. The radius is half of the diameter.

Geometry — Polygons

Number of Sides	Name	Number of Sides	Name
3 	triangle	7 	heptagon
4 	quadrilateral	8 	octagon
5 	pentagon	9 	nonagon
6 	hexagon	10 	decagon

Geometry — Triangles
equilateral — a triangle in which all 3 sides have the same length.
isosceles — a triangle in which 2 sides have the same length.
scalene — a triangle in which no sides are the same length.

Measurement — Relationships

Volume	**Distance**
3 teaspoons in a tablespoon	36 inches in a yard
2 cups in a pint	1,760 yards in a mile
2 pints in a quart	5,280 feet in a mile
4 quarts in a gallon	100 centimeters in a meter
Weight	1,000 millimeters in a meter
16 ounces in a pound	**Temperature**
2,000 pounds in a ton	0°Celsius - Freezing Point
Time	100°Celsius - Boiling Point
10 years in a decade	32°Fahrenheit - Freezing Point
100 years in a century	212°Fahrenheit - Boiling Point

Help Pages

Vocabulary (continued)

Ratio and Proportion

proportion — a statement that two ratios (or fractions) are equal. **Example**: $\frac{1}{2}=\frac{3}{6}$

ratio — a comparison of two numbers by division; a ratio looks like a fraction.
Example: $\frac{2}{5}$ or 2 to 5 or 2:5

percent (%) —the ratio of any number to 100. **Example**: 14% means 14 out of 100 or $\frac{14}{100}$.

Statistics

mean — the average of a group of numbers. The mean is found by finding the sum of a group of numbers and then dividing the sum by the number of members in the group.

Example: The average of 12, 18, 26, 17, and 22 is **19**. $\frac{12+18+26+17+22}{5}=\frac{95}{5}=19$

median — the middle value in a group of numbers. The median is found by listing the numbers in order from least to greatest, and finding the one that is in the middle of the list. If there is an even number of members in the group, the median is the average of the two middle numbers.

Example: The median of 14, 17, 24, 11, and 26 is **17**. 11, 14, (17), 24, 26

The median of 77, 93, 85, 95, 70, and 81 is **83**. 70, 77, (81, 85), 93, 95 $\frac{81+85}{2}=83$

mode — the number that occurs most often in a group of numbers. The mode is found by counting how many times each number occurs in the list. The number that occurs more than any other is the mode. Some groups of numbers have more than one mode.

Example: The mode of 77, (93), 85, (93), 77, 81, (93), and 71 is **93**. (93 occurs more than the others.)

Place Value

Whole Numbers

8,	9	6	3,	2	7	1,	4	0	5
Billions	Hundred Millions	Ten Millions	Millions	Hundred Thousands	Ten Thousands	Thousands	Hundreds	Tens	Ones

The number above is read: eight billion, nine hundred sixty-three million, two hundred seventy-one thousand, four hundred five.

Help Pages

Place Value

Decimal Numbers

1	7	8	.	6	4	0	5	9	2
Hundreds	Tens	Ones	*Decimal Point*	Tenths	Hundredths	Thousandths	Ten-thousandths	Hundred-thousandths	Millionths

The number above is read: one hundred seventy-eight and six hundred forty thousand, five hundred ninety-two millionths.

Solved Examples

Factors and Multiples

The **prime factorization** of a number is when a number is written as a product of its prime factors. A factor tree is helpful in finding the prime factors of a number.

Example: Use a factor tree to find the prime factors of 45.

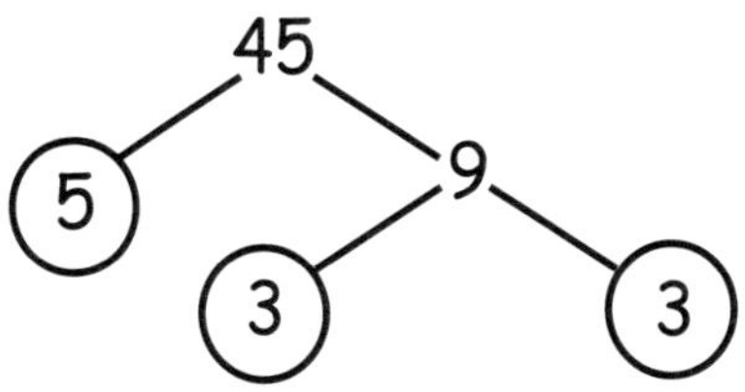

1. Find any 2 factors of 45 (5 and 9).
2. If a factor is prime, circle it. If a factor is not prime, find 2 factors of it.
3. Continue until all factors are prime.
4. In the final answer, the prime factors are listed in order, least to greatest, using exponents when needed.

The prime factorization of 45 is $3 \times 3 \times 5$ or $3^2 \times 5$.

The **greatest common factor (GCF)** is the largest factor that 2 numbers have in common.

Example: Find the greatest common factor (GCF) of 32 and 40.

The factors of 32 are 1, 2, 4, (8), 16, 32

The factors of 40 are 1, 2, 4, 5, (8), 10, 20, 40

The GCF of 32 and 40 is **8**.

1. First list the factors of each number.
2. Find the largest number that is in both lists.

Help Pages

Solved Examples

Factors and Multiples

The **least common multiple (LCM)** is the smallest multiple that two numbers have in common. The prime factors of the numbers can be useful in finding the LCM.

Example: Find the least common multiple (LCM) of 16 and 24.

```
2 | 16, 24
  2 | 8, 12
    2 | 4, 6
      2 | 2, 3
        3 | 1, 3
            1, 1
```

1. If any of the numbers are even, factor out a 2.
2. Continue factoring out 2 until all numbers left are odd.
3. If the prime number cannot be divided evenly into the number, simply bring the number down.
4. Once you are left with all 1's at the bottom, you're finished!
5. Multiply all of the prime numbers (on the left side of the bracket) together to find the least common multiple.

The LCM of 16 and 24 is
2 X 2 X 2 X 2 X 3 or **48**.

Fractions

Changing from an **improper fraction** to a **mixed number.**

Example: Change the improper fraction $\frac{5}{2}$ to a mixed number.

$\frac{5}{2}$ (five halves) means $5 \div 2$.

So, $\frac{5}{2}$ is equal to 2 wholes and 1 half or $2\frac{1}{2}$.

$$\begin{array}{r} 2 \text{ wholes} \\ 2\overline{)5} \\ -4 \\ \hline 1 \text{ half} \end{array}$$

Changing from a mixed number to an improper fraction.

Example: Change the mixed number $7\frac{1}{4}$ to an improper fraction.

1. You're going to make a new fraction. To find the numerator of the new fraction, multiply the whole number by the denominator, and add the numerator
2. Keep the same denominator in your new fraction as you had in the mixed number.

$7\frac{1}{4}$ 7 x 4 = 28. 28 + 1 = 29.

The new numerator is 29.

Keep the same denominator, 4.

The new fraction is $\frac{29}{4}$.

$7\frac{1}{4}$ is equal to $\frac{29}{4}$.

Help Pages

Solved Examples

Fractions (continued)

Equivalent fractions are two fractions that are equal to each other. Usually you will be finding a missing numerator or denominator.

Example: Find a fraction that is equivalent to $\frac{4}{5}$ and has a denominator of 35.

$$\frac{4}{5} \overset{\times 7}{=} \frac{?}{35}$$

1. Ask yourself, "What did I do to 5 to get 35?" (Multiply by 7.)
2. Whatever you did in the denominator, you also must do in the numerator. 4 x 7 = 28. The missing numerator is 28.

So, $\frac{4}{5}$ is equivalent to $\frac{28}{35}$.

Example: Find a fraction that is equivalent to $\frac{4}{5}$ and has a numerator of 24.

$$\frac{4}{5} \overset{\times 6}{=} \frac{24}{?}$$

1. Ask yourself, "What did I do to 4 to get 24?" (Multiply by 6.)
2. Whatever you did in the numerator, you also must do in the denominator. 5 x 6 = 30. The missing denominator is 30.

So, $\frac{4}{5}$ is equivalent to $\frac{24}{30}$.

Comparing fractions means looking at two or more fractions and determining if they are equal, if one is greater than (>) the other, or if one is less than (<) the other. A simple way to compare fractions is by cross-multiplying, using the steps below.

Examples: Compare these fractions. Use the correct symbol. $\frac{8}{9} \bigcirc \frac{3}{4}$ $\frac{7}{9} \bigcirc \frac{6}{7}$

$$\overset{32}{\frac{8}{9}} \; > \; \overset{27}{\frac{3}{4}}$$

$$\overset{49}{\frac{7}{9}} \; < \; \overset{54}{\frac{6}{7}}$$

So, $\frac{8}{9} > \frac{3}{4}$ and $\frac{7}{9} < \frac{6}{7}$

1. Begin with the denominator on the left and multiply by the opposite numerator. Put the answer (product) above the side where you ended. (9 x 3 = 27)
2. Cross-multiply the other denominator and numerator and put that product above where you ended.
3. Compare the two products and insert the correct symbol.

HINT: **Always** multiply diagonally **upwards**!

To **add (or subtract) fractions with the same denominator**, simply add (or subtract) the numerators, keeping the same denominator.

Examples: $\frac{3}{5} + \frac{1}{5} = \frac{4}{5}$

$\frac{8}{9} - \frac{1}{9} = \frac{7}{9}$

To **add mixed numbers**, follow a process similar to the one you used with fractions. If the sum is an improper fraction, be sure to simplify it.

Example:

$$\begin{array}{r} 1\frac{2}{5} \\ +1\frac{4}{5} \\ \hline 2\frac{6}{5} \end{array}$$

$2\frac{6}{5}$ is improper. $\frac{6}{5}$ can be rewritten as $1\frac{1}{5}$.

So, $2\frac{6}{5}$ is $2 + 1\frac{1}{5} = 3\frac{1}{5}$.

Help Pages

Solved Examples

Fractions (continued)

When **adding fractions that have different denominators**, you need to change the fractions so they have a common denominator before they can be added.

Finding the **least common denominator (LCD)**:

The LCD of the fractions is the same as the least common multiple of the denominators. Sometimes, the LCD will be the product of the denominators.

Example: Find the sum of $\frac{3}{8}$ and $\frac{1}{12}$.

$$\frac{3}{8} = \frac{9}{24}$$
$$+\frac{1}{12} = \frac{2}{24}$$
$$\frac{11}{24}$$

1. First, find the LCM of 8 and 12.
2. The LCM of 8 and 12 is 24. This is also the LCD of these 2 fractions.
3. Find an equivalent fraction for each that has a denominator of 24.
4. When they have a common denominator, the fractions can be added.

2	8, 12
2	4, 6
2	2, 3
3	1, 3
	1, 1

$2 \times 2 \times 2 \times 3 = 24$

The LCM is 24.

So, the LCD is 24.

Example: Add $\frac{1}{4}$ and $\frac{1}{5}$.

$$\frac{1}{4} = \frac{5}{20}$$
$$+\frac{1}{5} = \frac{4}{20}$$
$$\frac{9}{20}$$

$4 \times 5 = 20$ The LCD is 20.

When **adding mixed numbers with unlike denominators**, follow a process similar to the one you used with fractions (above). Be sure to put your answer in simplest form.

Example: Find the sum of $6\frac{3}{7}$ and $5\frac{2}{3}$.

$$6\frac{3}{7} = 6\frac{9}{21}$$
$$+5\frac{2}{3} = 5\frac{14}{21}$$
$$11\frac{23}{21}$$

improper

$$\frac{23}{21} = 1\frac{2}{21} + 11 = 12\frac{2}{21}$$

1. Find the LCD.
2. Find the missing numerators.
3. Add the whole numbers, then add the fractions.
4. Make sure your answer is in simplest form.

Help Pages

Solved Examples

Fractions (continued)

When **subtracting numbers with unlike denominators**, follow a process similar to the one you used when adding fractions. Be sure to put your answer in simplest form.

Examples: Find the difference of $\frac{3}{4}$ and $\frac{2}{5}$. Subtract $\frac{1}{16}$ from $\frac{3}{8}$.

$$\frac{3}{4} = \frac{15}{20}, \quad -\frac{2}{5} = \frac{8}{20}, \quad \frac{7}{20}$$

1. Find the LCD just as you did when adding fractions.
2. Find the missing numerators.
3. Subtract the numerators and keep the common denominator.
4. Make sure your answer is in simplest form.

$$\frac{3}{8} = \frac{6}{16}, \quad -\frac{1}{16} = \frac{1}{16}, \quad \frac{5}{16}$$

When **subtracting mixed numbers with unlike denominators**, follow a process similar to the one you used when adding mixed numbers. Be sure to put your answer in simplest form.

Example: Subtract $4\frac{2}{5}$ from $8\frac{9}{10}$.

1. Find the LCD.
2. Find the missing numerators.
3. Subtract and simplify your answer.

$$8\frac{9}{10} = 8\frac{9}{10}, \quad -4\frac{2}{5} = 4\frac{4}{10}, \quad 4\frac{5}{10} = 4\frac{1}{2}$$

Sometimes when subtracting mixed numbers you may need to rename them. If the numerator of the top fraction is smaller than the numerator of the bottom fraction, rename the mixed number as in the example below.

Example: Subtract $5\frac{5}{6}$ from $9\frac{1}{4}$.

1. Find the LCD.
2. Find the missing numerators. Write equivalent fractions.
3. Because you can't subtract 10 from 3, rename the whole number as a mixed number using the common denominator.
4. Add the two fractions to get an improper fraction.
5. Subtract the whole numbers and the fractions and simplify your answer.

$$9\frac{1}{4} = 9\frac{3}{12} = 8\frac{12}{12} + \frac{3}{12} = 8\frac{15}{12}$$
$$-5\frac{5}{6} = 5\frac{10}{12} = \quad 5\frac{10}{12}$$
$$3\frac{5}{12}$$

More examples:

$$8\frac{1}{2} = 8\frac{2}{4} = 7\frac{4}{4} + \frac{2}{4} = 7\frac{6}{4}$$
$$-4\frac{3}{4} = 4\frac{3}{4} = \quad -4\frac{3}{4}$$
$$3\frac{3}{4}$$

$$10\frac{1}{5} = 10\frac{4}{20} = 9\frac{20}{20} + \frac{4}{20} = 9\frac{24}{20}$$
$$-6\frac{3}{4} = 6\frac{15}{20} = \quad -6\frac{15}{20}$$
$$3\frac{9}{20}$$

Help Pages

Solved Examples

Fractions (continued)

To **multiply fractions**, simply multiply the numerators together to get the numerator of the product. Then multiply the denominators together to get the denominator of the product. Make sure your answer is in simplest form.

Examples: Multiply $\frac{3}{5}$ by $\frac{2}{3}$.

$$\frac{3}{5} \times \frac{2}{3} = \frac{6}{15} = \frac{2}{5}$$

1. Multiply the numerators.
2. Multiply the denominators.
3. Simplify your answer.

Multiply $\frac{5}{8}$ by $\frac{4}{5}$.

$$\frac{5}{8} \times \frac{4}{5} = \frac{20}{40} = \frac{1}{2}$$

Sometimes you can use cancelling when multiplying fractions. Let's look at the examples again.

$$\frac{\cancel{3}^{1}}{5} \times \frac{2}{\cancel{3}_{1}} = \frac{2}{5}$$

The 3s have a common factor — 3. Divide both of them by 3. Since, 3 ÷ 3 = 1, we cross out the 3s and write 1s in their place.

Now, multiply the fractions. In the numerator, 1 × 2 = 2. In the denominator, 5 × 1 = 5 .

The answer is $\frac{2}{5}$.

1. Are there any numbers in the numerator and the denominator that have common factors?
2. If so, cross out the numbers, divide both by that factor, and write the quotient.
3. Then, multiply the fractions as described above, using the quotients instead of the original numbers.

$$\frac{\cancel{5}^{1}}{{}_{2}\cancel{8}} \times \frac{\cancel{4}^{1}}{\cancel{5}_{1}} = \frac{1}{2}$$

As in the other example, the 5s can be cancelled.

But here, the 4 and the 8 also have a common factor — 4.

8 ÷ 4 = 2 and 4 ÷ 4 = 1.

After cancelling both of these, you can multiply the fractions.

REMEMBER: You can cancel up and down or diagonally, but NEVER sideways!

When **multiplying mixed numbers**, you must first change them into improper fractions.

Examples: Multiply $2\frac{1}{4}$ by $3\frac{1}{9}$.

$$2\frac{1}{4} \times 3\frac{1}{9} =$$

$$\frac{\cancel{9}^{1}}{{}_{1}\cancel{4}} \times \frac{\cancel{28}^{7}}{\cancel{9}_{1}} = \frac{7}{1} = 7$$

1. Change each mixed number to an improper fraction.
2. Cancel wherever you can.
3. Multiply the fractions.
4. Put your answer in simplest form.

Multiply $3\frac{1}{8}$ by 4.

$$3\frac{1}{8} \times 4 =$$

$$\frac{25}{{}_{2}\cancel{8}} \times \frac{\cancel{4}^{1}}{1} = \frac{25}{2} = 12\frac{1}{2}$$

To **divide fractions**, you must take the reciprocal of the 2nd fraction, and then multiply that reciprocal by the 1st fraction. Don't forget to simplify your answer!

Examples: Divide $\frac{1}{2}$ by $\frac{7}{12}$.

$$\frac{1}{2} \div \frac{7}{12} =$$

$$\frac{1}{{}_{1}\cancel{2}} \times \frac{\cancel{12}^{6}}{7} = \frac{6}{7}$$

1. Keep the 1st fraction as it is.
2. Write the reciprocal of the 2nd fraction.
3. Change the sign to multiplication.
4. Cancel if you can and multiply.
5. Simplify your answer.

Divide $\frac{7}{8}$ by $\frac{3}{4}$.

$$\frac{7}{8} \div \frac{3}{4} =$$

$$\frac{7}{{}_{2}\cancel{8}} \times \frac{\cancel{4}^{1}}{3} = \frac{7}{6} = 1\frac{1}{6}$$

Help Pages

Solved Examples

Fractions (continued)

When **dividing mixed numbers**, you must first change them into improper fractions.

Example: Divide $1\frac{1}{4}$ by $3\frac{1}{2}$.

$$1\frac{1}{4} \div 3\frac{1}{2} =$$

$$\frac{5}{4} \div \frac{7}{2} =$$

$$\frac{5}{\cancel{4}_{2}} \times \frac{\cancel{2}^{1}}{7} = \frac{5}{14}$$

1. Change each mixed number to an improper fraction.
2. Keep the 1st fraction as it is.
3. Write the reciprocal of the 2nd fraction.
4. Change the sign to multiplication.
5. Cancel if you can and multiply.
6. Simplify your answer.

Decimals

When we **compare decimals**, we are looking at two or more decimal numbers and deciding which has the smaller or larger value. We sometimes compare by placing them in order from least to greatest or from greatest to least. Another way to compare is to use the symbols for "less than" (<), "greater than" (>), or "equal to" (=).

Example: Order these numbers from least to greatest. 0.561 0.506 0.165

1. Write the numbers in a column, lining up the decimal points.
2. Write zeroes, if necessary, so all have the same number of digits.
3. Begin on the left and compare the digits.

0.561
0.506
0.165

Since they all have 3 digits, we don't need to add zeroes.

Beginning on the left, the fives are equal, but the one is less, so 0.165 is the smallest.

Then, look at the next digit. The zero is less than the six, so 0.506 is next smallest.

So, in order from least to greatest:

0.165, 0.506, 0.561

Example: Place these numbers in order from greatest to least. 0.44 0.463 0.045

0.440
0.463
0.045

After lining up the numbers, we must add a zero to 0.44 to make them all have the same number of digits.

Beginning on the left, the zero is smaller than the fours, so 0.045 is the smallest.

Look at the next digit. The four is smaller than the six, so 0.440 is the next smallest.

In order from greatest to least: 0.463, 0.440, 0.045

Help Pages

Solved Examples

Decimals (continued)

When we **round decimals**, we are approximating them. This means we end the decimal at a certain place value and we decide if it's closer to the next higher number (round up) or to the next lower number (keep the same). It might be helpful to look at the decimal place-value chart on p. 287.

Example: Round 0.574 to the tenths place.

There is a 5 in the rounding (tenths) place. → $0.\underline{5}74$

Since 7 is greater than 5, change the 5 to a 6. → $0.\underline{5}74$

Drop the digits to the right of the tenths place. → 0.6

1. Identify the number in the rounding place.
2. Look at the digit to its right.
3. If the digit is 5 or greater, increase the number in the rounding place by 1. If the digit is less than 5, keep the number in the rounding place the same.
4. Drop all digits to the right of the rounding place.

Example: Round 2.783 to the nearest hundredth.

$2.7\underline{8}3$ ← There is an 8 in the rounding place.

$2.7\underline{8}3$ ← Since 3 is less than 5, keep the rounding place the same.

2.78 ← Drop the digits to the right of the hundredths place.

Adding and subtracting decimals is very similar to adding or subtracting whole numbers. The main difference is that you have to line-up the decimal points in the numbers before you begin.

Examples: Find the sum of 3.14 and 1.2.

$$\begin{array}{r} 3.14 \\ +\ 1.20 \\ \hline 4.34 \end{array}$$

Add 55.1, 6.472, and 18.33.

$$\begin{array}{r} 55.100 \\ 6.472 \\ +18.330 \\ \hline 79.902 \end{array}$$

1. Line up the decimal points. Add zeroes as needed.
2. Add (or subtract) the decimals.
3. Add (or subtract) the whole numbers.
4. Bring the decimal point straight down.

Examples: Subtract 3.7 from 9.3.

$$\begin{array}{r} 9.3 \\ -\ 3.7 \\ \hline 5.6 \end{array}$$

Find the difference of 4.1 and 2.88.

$$\begin{array}{r} 4.10 \\ -2.88 \\ \hline 1.22 \end{array}$$

Help Pages

Solved Examples

Decimals (continued)

When **multiplying a decimal by a whole number**, the process is similar to multiplying whole numbers.

Examples: Multiply 3.42 by 4.

3.42 → 2 decimal places
× 4 → 0 decimal places
13.68 → Place decimal point so there are 2 decimal places.

1. Line up the numbers on the right.
2. Multiply. Ignore the decimal point.
3. Place the decimal point in the product. (The total number of decimal places in the product must equal the total number of decimal places in the factors.)

Find the product of 2.3 and 2.

2.3 → 1 decimal place
× 2 → 0 decimal places
4.6 → Place decimal point so there is 1 decimal place.

The process for **multiplying two decimal numbers** is a lot like what we just did above.

Examples: Multiply 0.4 by 0.6.

0.4 → 1 decimal place
× 0.6 → 1 decimal place
0.24 → Place decimal point so there are 2 decimal places.

Find the product of 2.67 and 0.3.

2.67 → 2 decimal places
× 0.3 → 1 decimal place
0.801 → Place decimal point so there are 3 decimal places.

Sometimes it is necessary to add **zeroes in the product** as placeholders in order to have the correct number of decimal places.

Example: Multiply 0.03 by 0.4.

0.03 → 2 decimal places
× 0.4 → 1 decimal place
0.012 → Place decimal point so there are 3 decimal places.

We had to add a zero in front of the 12 so that we could have 3 decimal places in the product.

The process for **dividing a decimal number by a whole number** is similar to dividing whole numbers.

Examples: Divide 6.4 by 8.

$$\begin{array}{r} 0.8 \\ 8\overline{)6.4} \\ -6\ 4 \\ \hline 0 \end{array}$$

1. Set up the problem for long division.
2. Place the decimal point in the quotient directly above the decimal point in the dividend.
3. Divide. Add zeroes as placeholders if necessary. (See examples below.)

Find the quotient of 20.7 and 3.

$$\begin{array}{r} 6.9 \\ 3\overline{)20.7} \\ -18 \\ \hline 27 \\ -27 \\ \hline 0 \end{array}$$

Examples: Divide 4.5 by 6.

$$\begin{array}{r} 0.75 \\ 6\overline{)4.50} \\ -42\downarrow \\ \hline 30 \\ -3\ 0 \\ \hline 0 \end{array}$$

Add zero(es).

Bring zero down.

Keep dividing.

Find the quotient of 3.5 and 4.

$$\begin{array}{r} 0.875 \\ 4\overline{)3.500} \\ -3\ 2\downarrow\downarrow \\ \hline 30\downarrow \\ -28\downarrow \\ \hline 20 \\ -20 \\ \hline 0 \end{array}$$

Help Pages

Solved Examples

Decimals (continued)

When dividing decimals the remainder is not always zero. Sometimes, the division continues on and on and the remainder begins to repeat itself. When this happens the quotient is called a **repeating decimal**.

Examples: Divide 2 by 3.

$$3\overline{)2.000} = 0.\overline{6}6$$

```
   0.66
3 )2.000
 - 18↓↓
    20↓
  - 18↓
     20
```

Divide 10 by 11.

```
     0.9090
11 )10.00000
  - 99↓↓↓↓
     100↓↓
   -  99↓↓
       100
```

Add zeroes as needed

This pattern (with the same remainder begins to repeat itself.

To write the final answer, put a bar in the quotient over the digits that repeat.

The process for **dividing a decimal number by a decimal number** is similar to other long division that you have done. The main difference is that we have to move the decimal point in both the dividend and the divisor the <u>same number of places</u> to the right.

1. Change the divisor to a whole number by moving the decimal point as many places to the right as needed.
2. Move the decimal in the dividend the same number of places to the right as you did in the divisor.
3. Put the decimal point in the quotient directly above the decimal point in the dividend.
4. Divide.

Examples: Divide 1.8 by 0.3.

```
      6.
0.3 )1.8
    -18
      0
```

Divide 0.385 by 0.05.

```
         7.7
0.05 )0.385
      -35↓
        35
       -35
         0
```

Geometry

Finding the **area of a parallelogram** is similar to finding the area of any other quadrilateral. The area of the figure is equal to the length of its base multiplied by the height of the figure.

Area of parallelogram = base × height   or $A = b \times h$

Example: Find the area of the parallelogram below.

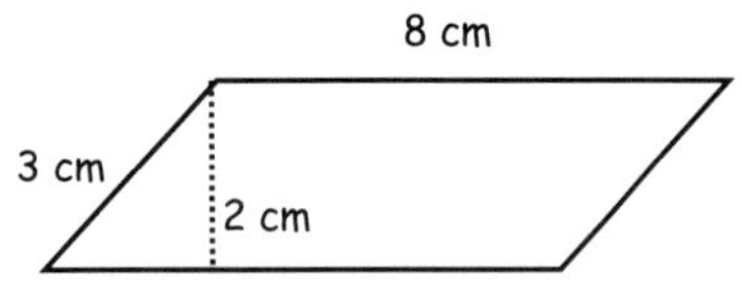

1. Find the length of the base. (8 cm)
2. Find the height. (It is 2 cm. The height is always straight up and down – never slanted.)
3. Multiply to find the area. (16 cm^2)

So, A = 8 cm × 2 cm = 16 cm^2.

Help Pages

Solved Examples

Geometry (continued)

To find the **area of a triangle**, it is helpful to recognize that any triangle is exactly half of a parallelogram.

The whole figure is a parallelogram.

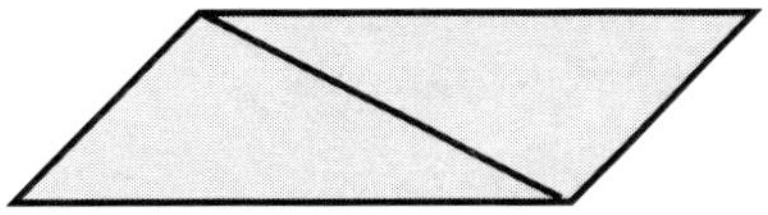

Half of the whole figure is a triangle.

So, the triangle's area is equal to half of the product of the base and the height.

$$\text{Area of triangle} = \frac{1}{2}(\text{base} \times \text{height}) \quad \text{or} \quad A = \frac{1}{2}bh \quad \text{or} \quad A = \frac{bh}{2}$$

Examples: Find the area of the triangles below.

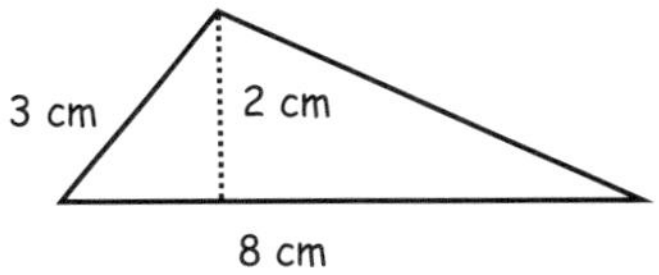

1. Find the length of the base. (8 cm)
2. Find the height. (It is 2 cm. The height is always straight up and down – never slanted.)
3. Multiply them together and divide by 2 to find the area. (8 cm^2)

So, $A = 8 \text{ cm} \times 2 \text{ cm} \times \frac{1}{2} = 8 \text{ cm}^2$.

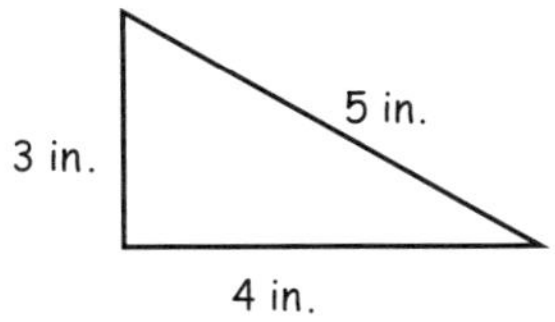

The base of this triangle is 4 inches long. Its height is 3 inches. (Remember the height is always straight up and down!)

So, $A = 4 \text{ in.} \times 3 \text{ in.} \times \frac{1}{2} = 6 \text{ in}^2$.

Finding the **area of a trapezoid** is a little different than the other quadrilaterals that we have seen. Trapezoids have 2 bases of unequal length. To find the area, first find the average of the lengths of the 2 bases. Then, multiply that average by the height.

$$\text{Area of trapezoid} = \frac{\text{base}_1 + \text{base}_2}{2} \times \text{height} \quad \text{or} \quad A = \left(\frac{b_1 + b_2}{2}\right) \times h$$

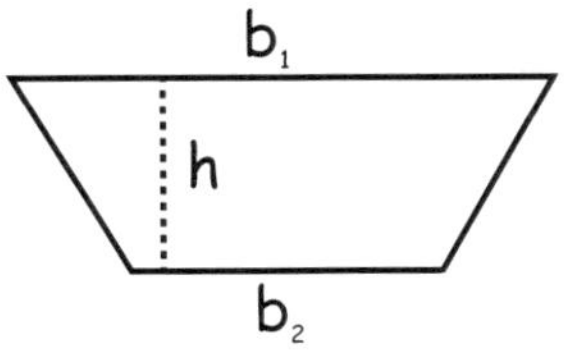

The bases are labeled b_1 and b_2.

The height, h, is the distance between the bases.

Example: Find the area of the trapezoid below.

1. Add the lengths of the two bases. (22 cm)
2. Divide the sum by 2. (11 cm)
3. Multiply that result by the height to find the area. (110 cm^2)

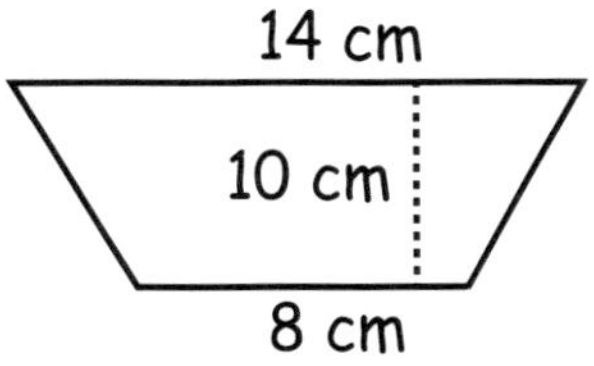

$$\frac{14 \text{ cm} + 8 \text{ cm}}{2} = \frac{22 \text{ cm}}{2} = 11 \text{ cm}$$

$11 \text{ cm} \times 10 \text{ cm} = \mathbf{110 \text{ cm}^2} = \text{Area}$

Help Pages

Solved Examples

Geometry (continued)

The **circumference of a circle** is the distance around the outside of the circle. Before you can find the circumference of a circle you must know either its radius or its diameter. Also, you must know the value of the constant, pi (π). $\pi = 3.14$ (rounded to the nearest hundredth)

Once you have this information, the circumference can be found by multiplying the diameter by pi.

Circumference = $\pi \times$ diameter or $C = \pi d$

Examples: Find the circumference of the circles below.

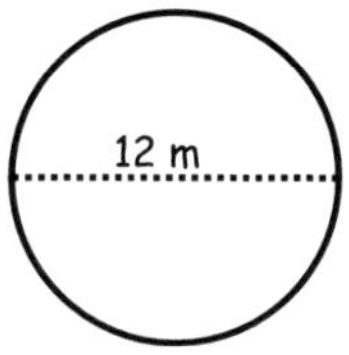

1. Find the length of the diameter. (12 m)
2. Multiply the diameter by π. (12 m × 3.14)
3. The product is the circumference. (37.68 m)

So, C = 12 m × 3.14 = 37.68 m.

Sometimes the radius of a circle is given instead of the diameter. Remember, the radius of any circle is exactly half of the diameter. If a circle has a radius of 3 feet, its diameter is 6 feet.

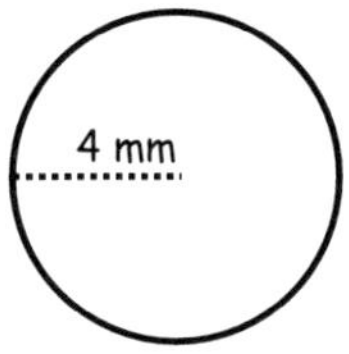

1. Since the radius is 4 mm, the diameter must be 8 mm.
2. Multiply the diameter by π. (8 mm × 3.14)
3. The product is the circumference. (25.12 mm)

So, C = 8 mm × 3.14 = 25.12 mm.

When finding the **area of a circle**, the length of the radius is squared (multiplied by itself), and then that answer is multiplied by the constant, pi (π). π = 3.14 (rounded to the nearest hundredth).

Area = π × radius × radius or $A = \pi r^2$

Examples: Find the area of the circles below.

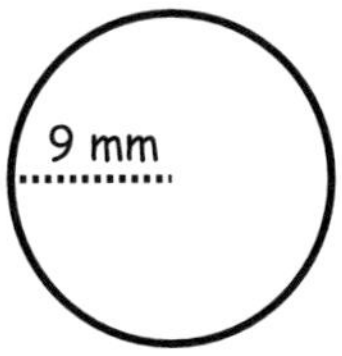

1. Find the length of the radius. (9 mm)
2. Multiply the radius by itself. (9 mm x 9 mm)
3. Multiply the product by pi. (81 mm^2 x 3.14)
4. The result is the area. (254.34 mm^2)

So, A = 9 mm x 9 mm x 3.14 = **254.34 mm^2**.

Sometimes the diameter of a circle is given instead of the radius. Remember, the diameter of any circle is exactly twice the radius. If a circle has a diameter of 6 feet, its radius is 3 feet.

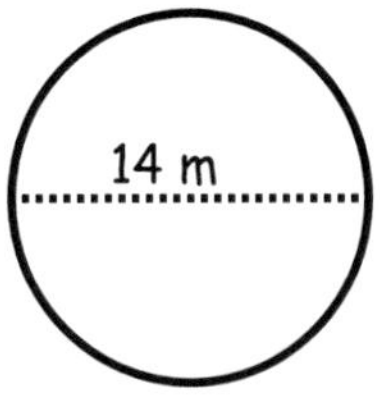

1. Since the diameter is 14 m, the radius must be 7 m.
2. Square the radius. (7 m x 7 m)
3. Multiply that result by π. (49 m^2 × 3.14).
4. The product is the area. (153.86 m^2)

So, A = (7 m)^2 x 3.14 = **153.86 m^2**.

Help Pages

Solved Examples

Geometry (continued)

To find the **surface area** of a solid figure, it is necessary to first count the total number of faces. Then, find the area of each of the faces; finally, add the areas of each face. That sum is the surface area of the figure.

Here, the focus will be on finding the surface area of a rectangular prism. A rectangular prism has 6 faces. Actually, the opposite faces are identical, so this figure has 3 pairs of faces. Also, a prism has only 3 dimensions: length, width, and height.

This prism has identical left & right sides (A & B), identical top and bottom (C & D), and identical front and back (unlabeled).

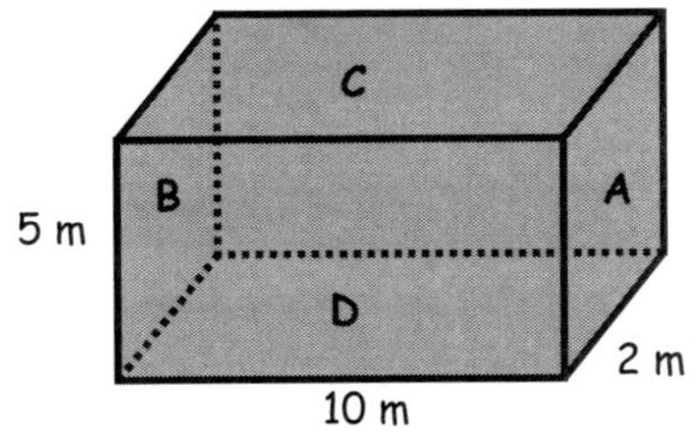

1. Find the area of the front: $l \times w$. ($10 \text{ m} \times 5 \text{ m} = 50 \text{ m}^2$)
   Since the back is identical, its area is the same.
2. Find the area of the top (C): $l \times h$. ($10 \text{ m} \times 2 \text{ m} = 20 \text{ m}^2$)
   Since the bottom (D) is identical, its area is the same.
3. Find the area of side A: $w \times h$. ($2 \text{ m} \times 5 \text{ m} = 10 \text{ m}^2$)
   Sinceside B is identical, its area is the same.
4. Add up the areas of all 6 faces.
   ($10 \text{ m}^2 + 10 \text{ m}^2 + 20 \text{ m}^2 + 20 \text{ m}^2 + 50 \text{ m}^2 + 50 \text{ m}^2 = 160 \text{ m}^2$)

The formula is Surface Area = 2(length x width) + 2(length x height) + 2(width x height)
or $SA = 2lw + 2lh + 2wh$

Ordering Integers

Integers include the counting numbers, their opposites (negative numbers) and zero.

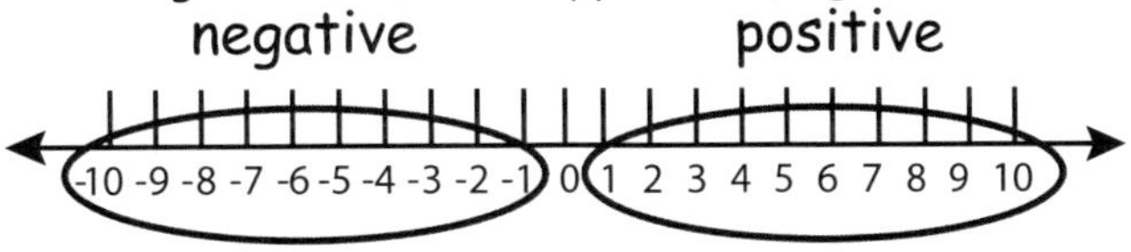

The negative numbers are to the left of zero. The positive numbers are to the right of zero.

The further a number is to the right, the greater its value. For example, 9 is further to the right than 2, so 9 is greater than 2.

In the same way, -1 is further to the right than -7, so -1 is greater than -7.

Examples: Order these integers from **least to greatest**: -10, 9, -25, 36, 0

Remember, the smallest number will be the one farthest to the left on the number line, -25, then -10, then 0. Next will be 9, and finally 36.

Answer: -25, -10, 0, 9, 36

Order these integers from **least to greatest**: -44, -19, -56, -80, -2
Answer: -80, -56, -44, -19, -2 (-80 is farthest to the left, so it is smallest. -2 is farthest to the right, so it is greatest.)

Put these integers in order from **greatest to least**: -94, -6, -24, -70, -14
Now the greatest value (farthest to the right) will come first and the smallest value (farthest to the left) will come last.
Answer: -6, -14, -24, -70, -94

Help Pages

Solved Examples

Ratio and Proportion

A **ratio** is used to compare two numbers. There are three ways to write a ratio comparing 5 and 7.

1. Word form ➡ 5 to 7
2. Fraction form ➡ $\frac{5}{7}$
3. Ratio form ➡ 5 : 7

All are read as " five to seven."

You must make sure that all ratios are written in simplest form. (Just like fractions!!)

A **proportion** is a statement showing that two ratios are equal to each other. There are two ways to solve a proportion when a number is missing.

1. One way to solve a proportion is already familiar to you. You can use the equivalent fraction method.

$$\frac{5}{8} = \frac{n}{64} \quad (\times 8)$$

$n = 40$

So, $\frac{5}{8} = \frac{40}{64}$.

To use Cross-Products:

1. Multiply downward on each diagonal.
2. Make the product of each diagonal equal to each other.
3. Solve for the missing variable.

2. Another way to solve a proportion is by using cross-products.

$$\frac{14}{20} = \frac{21}{n}$$

$20 \times 21 = 14 \times n$

$420 = 14n$

$\frac{420}{14} = \frac{14n}{14}$

$30 = n$

So, $\frac{14}{20} = \frac{21}{30}$.

Percent

When changing from a fraction to a percent, a decimal to a percent, or from a percent to either a fraction or a decimal, it is very helpful to use an FDP chart (Fraction, Decimal, Percent).

To change a **fraction to a percent and/or decimal**, first find an equivalent fraction with 100 in the denominator. Once you have found that equivalent fraction, it can easily be written as a decimal. To change that decimal to a percent, move the decimal point 2 places to the right and add a % sign.

Example: Change $\frac{2}{5}$ to a percent and then to a decimal.

1. Find an equivalent fraction with 100 in the denominator.
2. From the equivalent fraction above, the decimal can easily be found. Say the name of the fraction: "forty hundredths." Write this as a decimal: 0.40.
3. To change 0.40 to a percent, move the decimal two places to the right. Add a % sign.

F	D	P
$\frac{2}{5}$		

F	D	P
$\frac{2}{5} = \frac{?}{100}$	0.40	

F	D	P
$\frac{2}{5} = \frac{?}{100}$	0.40	40%

$$\frac{2}{5} = \frac{?}{100} \quad (\times 20)$$

$? = 40$

$\frac{2}{5} = \frac{40}{100} = 0.40$

$0.40 = 40\%$

Help Pages

Solved Examples

Percent

When changing from a **percent to a decimal or a fraction**, the process is similar to the one used on the previous page. Begin with the percent. Write it as a fraction with a denominator of 100; reduce this fraction. Return to the percent, move the decimal point 2 places to the left. This is the decimal.

Example: Write 45% as a fraction and then as a decimal.

1. Begin with the percent. (45%) Write a fraction where the denominator is 100 and the numerator is the "percent." $\frac{45}{100}$
2. This fraction must be reduced. The reduced fraction is $\frac{9}{20}$.
3. Go back to the percent. Move the decimal point two places to the left to change it to a decimal.

F	D	P
$\frac{45(\div 5)}{100(\div 5)} = \frac{9}{20}$		45%

F	D	P
$\frac{9}{20}$		45% = 0.45

F	D	P
$\frac{9}{20}$	0.45	45%

When changing from a **decimal to a percent or a fraction**, again, the process is similar to the one used above. Begin with the decimal. Move the decimal point 2 places to the right and add a % sign. Return to the decimal. Write it as a fraction and reduce.

Example: Write 0.12 as a percent and then as a fraction.

1. Begin with the decimal. (0.12) Move the decimal point two places to the right to change it to a percent.
2. Go back to the decimal and write it as a fraction. Reduce this fraction.

F	D	P
	0.12	

F	D	P
	0.12 = 12%	12%

F	D	P
$\frac{12(\div 4)}{100(\div 4)} = \frac{3}{25}$	0.12	12%

Help Pages

Solved Examples

Compound Probability

The **probability of two or more independent events** occurring together can be determined by multiplying the individual probabilities together. The product is called the compound probability.

Probability of A & B = (Probability of A) x (Probability of B)

or $P(A \text{ and } B) = P(A) \times P(B)$

Example: What is the probability of rolling a 6 and then a 2 on two rolls of a die [P(6 and 2)]?

A) First, find the probability of rolling a 6 [P(6)]. Since there are 6 numbers on a die and only one of them is a 6, the probability of getting a 6 is $\frac{1}{6}$.

B) Then find the probability of rolling a 2 [P(2)]. Since there are 6 numbers on a die and only one of them is a 2, the probability of getting a 2 is $\frac{1}{6}$.

So, $P(6 \text{ and } 2) = P(6) \times P(2) = \frac{1}{6} \times \frac{1}{6} = \frac{1}{36}$

There is a 1 to 36 chance of getting a 6 and then a 2 on two rolls of a die.

Example: What is the probability of getting a 4 and then a number greater than 2 on two spins of this spinner [P(4 and greater than 2)]?

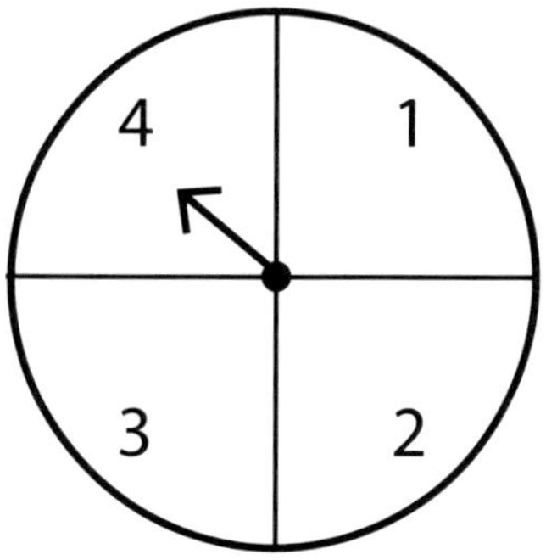

A) First, find the probability of getting a 4 [P(4)]. Since there are 4 numbers on the spinner and only one of them is a 4, the probability of getting a 4 is $\frac{1}{4}$.

B) Then find the probability of getting a number greater than 2 [P(greater than 2)]. Since there are 4 numbers on the spinner and two of them are greater than 2, the probability of getting a 2 is $\frac{2}{4}$.

So, $P(2 \text{ and greater than } 2) = P(2) \times P(\text{greater than } 2) = \frac{1}{4} \times \frac{2}{4} = \frac{2}{16} = \frac{1}{8}$

There is a 1 to 8 chance of getting a 4 and then a number greater than 2 on two spins of a spinner.

Example: On three flips of a coin, what is the probability of getting heads, tails, heads [P(H,T,H)]?

A) First, find the probability of getting heads [P(H)]. Since there are only 2 sides on a coin and only one of them is heads, the probability of getting heads is $\frac{1}{2}$.

B) Then find the probability of getting tails [P(T)]. Again, there are only 2 sides on a coin and only one of them is tails. The probability of getting tails is also $\frac{1}{2}$.

So, $P(H,T,H) = P(H) \times P(T) \times P(H) = \frac{1}{2} \times \frac{1}{2} \times \frac{1}{2} = \frac{1}{8}$

There is a 1 to 8 chance of getting heads, tails and then heads on 3 flips of a coin.

Who Knows???

Degrees in a right angle? ....(90)

A straight angle? ....(180)

Angle greater than 90°? .... (obtuse)

Less than 90°? ....(acute)

Sides in a quadrilateral? ....(4)

Sides in an octagon?........................... (8)

Sides in a hexagon? ....(6)

Sides in a pentagon? .... (5)

Sides in a heptagon? ....(7)

Sides in a nonagon? .... (9)

Sides in a decagon? .... (10)

Inches in a yard? ....(36)

Yards in a mile? ....(1,760)

Feet in a mile? ....(5,280)

Centimeters in a meter? ....(100)

Teaspoons in a tablespoon? .... (3)

Ounces in a pound? ....(16)

Pounds in a ton?............................(2,000)

Cups in a pint? .... (2)

Pints in a quart? .... (2)

Quarts in a gallon? ....(4)

Millimeters in a meter? .... (1,000)

Years in a century? ....(100)

Years in a decade? ....(10)

Celsius freezing? ....(0°C)

Celsius boiling? ....(100°C)

Fahrenheit freezing? ....(32°F)

Fahrenheit boiling? ....(212°F)

Number with only 2 factors? .... (prime)

Perimeter? ....(add the sides)

Area? ....(length x width)

Volume? .... (length x width x height)

Area of parallelogram?..........................
.... (base x height)

Area of triangle? ....($\frac{1}{2}$ base x height)

Area of trapezoid?..................................
...............................($\frac{\text{base} + \text{base}}{2} \times \text{height}$)

Surface area of a rectangular prism ....2(LW) + 2(WH) + 2(LH

Area of a circle? ....(πr^2)

Circumference of a circle? ....(πd)

Triangle with no sides equal? ....
.... (scalene)

Triangle with 3 sides equal?...................
....(equilateral)

Triangle with 2 sides equal? ....
.... (isosceles)

Distance across the middle of a circle?
.... (diameter)

Half of the diameter? .... (radius)

Figures with the same size and shape? .... (congruent)

Figures with same shape, different sizes? ....(similar)

Number occurring most often? ....
.... (mode)

Middle number? ....(median)

Answer in addition? ....(sum)

Answer in division? ....(quotient)

Answer in multiplication? .... (product)

Answer in subtraction? ......(difference)